THEORIES OF ENGINEERING EXPERIMENTATION

Third Edition

Hilbert Schenck

Professor of Mechanical Engineering
University of Rhode Island

With a contribution by

Roger J. Hawks

Associate Professor of Aeronautical
and Mechanical Engineering
Tri-State University

HEMISPHERE PUBLISHING CORPORATION

Washington   London

McGRAW–HILL BOOK COMPANY

New York   St. Louis   San Francisco   Auckland   Bogotá
Düsseldorf   Johannesburg   London   Madrid   Mexico
Montreal   New Delhi   Panama   Paris   São Paulo
Singapore   Sydney   Tokyo   Toronto

THEORIES OF ENGINEERING EXPERIMENTATION

2 3 4 5 6 7 8 9 0   K P K P   7 8 3 2 1 0 9

This book was set in Press Roman by Hemisphere Publishing Corporation. The editors were Rolfe W. Larson and Lynne Lackenbach; the production supervisor was Rebekah McKinney; and the typesetter was Sandra F. Watts.
Kingsport Press, Inc., was printer and binder.

Library of Congress Cataloging in Publication Data

Schenck, Hilbert van Nydeck, date
  Theories of engineering experimentation.

  Bibliography: p.
  Includes index.
  1. Engineering–Laboratory manuals. 2. Engineering mathematics. I. Hawks, Roger, joint author. II. Title.
TA152.S3 1979   620'.001'51   78-12190
ISBN 0-07-055267-3

CONTENTS

PREFACE

This text and reference book is intended for the undergraduate mechanical, electrical, or civil engineering student to accompany laboratory course work and also to serve in the planning of senior projects and masters-level research in engineering. In this third edition, certain changes have been made to fit the book to the present situation in engineering education: increasing enrollments, cutbacks in laboratory hours and supervision, and more and more emphasis on projects and individual activities by upper division students. Less background is expected of the reader, and the text will self-teach a variety of topics with greater detail and explanation. For example, the whole central topic of graphical and mathematical analysis of test data has been expanded so that a student with no background in this area will be able to start at the beginning and transform a line of points into either an equation or a statistical number.

Although it is now impossible in a book of any reasonable length to deal with all the gear, ideas, and instrumentation that may be found in a college engineering lab today, it seems unlikely that a given undergraduate experiment would not call up one or more topics explained in the book, and in most cases, several would be useful as the test proceeds. As in past editions, the order of topics is intended to match the sequential character of an engineering experiment as closely as possible. The first three chapters deal mainly with instrument error and its propagation, and Chap. 4 with test planning as related to variable reduction using dimensional analysis.

In Chap. 5 the response and loading of instruments is explained, and in Chap. 6 the actual spacing of the test points is considered. The final three chapters deal with the various basic methods of data analysis and reduction and the establishment of functions and statistical conclusions.

Three main changes mark this edition. First, the in-text examples have been expanded, and all examples showing data are "real"; that is, they result from actual student tests run in college laboratories. This approach has a very significant advantage over examples that are "cooked" with made-up data. Students not only can see how the text material is applied to data but also how real data act and what can be inferred from them. For example, in Chap. 2, there are three sets of "random" data used in examples, having 350, 69, and 11 items. Thus when a student has completed this chapter, he or she not only will have seen the ideas applied to real data but also will have acquired a good idea of what can be done with samples of various sizes as far as normal-curve statistics are concerned. Also, in each case where an extensive in-text example has been given, similar data will be found in the problems at the end of the chapter to enable the reader to duplicate the analysis shown. In other words, the text will allow considerable self-teaching if the user wishes.

The second major improvement in the book consists of the addition of Chap. 5 on instrument loading and response, written for this edition by Professor Roger J. Hawks. This chapter, filled with solid, analytic approaches to instrument application, will enable the reader actually to estimate loading and response errors for many of the tools found in undergraduate labs.

The third major change relates to the chapters on data analysis. After examining 25 years worth of undergraduate lab reports, masters theses, and doctoral projects, it has become evident that virtually every college-level effort in engineering lab or research could fit its "output" into one of three simple classifications: data that plot a smooth curve on some sort of coordinate system, data that show some kind of graphic presentation but that are so scattered as to require some statistical intervention (in this book either least-squares or correlation), and data that have no graphical sense but require "pure" statistical reasoning to understand. The three final chapters in the book correspond to this scheme. By understanding this material the reader should be able to make analysis decisions far more simply and quickly than by following other possible arrangements for analysis of data. Again, the expanded use of real examples will aid the beginner in this search. A student need only plot the data (if it has an X-Y sense), then compare the resulting picture with the several presentations in Chaps. 7 and 8 to become "located" in this "hierarchy of precision."

We engineers in education face a promising but uncertain future. For better or worse there is no turning back to a simpler day; there are too many people in the world. There is no one else coming from the universities and colleges to take us safely through the coming years of shortage and resource crisis but our own students. If the lights are going to stay on, these young people have got to be more careful, efficient, and clever than we have been, and thoughtful, well-planned experiments will be the core of our students' crucial efforts.

THEORIES
OF ENGINEERING
EXPERIMENTATION

EXPERIMENTATION AS A SUBJECT
FOR STUDY

The common link among engineers, physical researchers, and social scientists is the experiment and experimentation. Biologists testing new drug compounds on living animals, physicists probing the tiny nucleus with huge pieces of equipment, engineers comparing several production methods, all follow very diverse paths; yet all run experiments. Furthermore, there are basic similarities in their experimental methods. They try to *control* their experiments or, as we shall say, *eliminate the effect of extraneous variables.* They are all concerned with the *accuracy of their instruments and data taking.* Each experimenter wishes to *reduce the number of variables* in any given test, for this means both speed and economy in his or her work. No matter how simple the test, the *testing sequence should be planned* before the start. Once a test is under way, it is hard to imagine a situation in which the *detection of malfunctions* would not be important, provided that we think of malfunction in its broadest sense. Related to this last point is *testing for reasonableness of the results.* Finally, every experiment requires *result analysis and interpretation,* for without this crucial step, the entire procedure is meaningless.

Now it is perfectly true that the emphasis on these particular topics will vary from one scientific field to another. Biologists, agricultural researchers, and social scientists are likely to think more in terms of pretest planning and statistical inference than physicists or engineers. The experimental systems of the former are often living plants or animals, each with its unique and uncomputable differences. Thus, many experimental variables are beyond the investigators' control, and they must somehow plan tests to minimize or eliminate these extraneous influences. Furthermore, biological experimenters will repeat tests over and over again, whereas physical experimenters may be satisfied with only a few readings. But this does not mean that sequence planning and statistical inference are of no use in engineering, for technological fields have their own special uses for such techniques. On the other hand, engineering experimenters, with their relatively exact data, are more likely to utilize graphs and formulas to express this new-found data, whereas biologists must often be content with simple tables or statistical figures. But we can make no hard and fast distinction. In industrial production-line experiments where the worker and machine interact, the boundaries between the two types of experimental approaches dissolve, and engineers must often resort to methods pioneered by persons in much removed fields of study.

1-1 THE ENGINEERING EXPERIMENT

We shall consider the engineering experiment in this book and, specifically, tests involving mechanical, materials-testing, fluid-flow, and electrical phenomena. We shall study such tests by using examples and methods from many areas of scientific thought, for through generality we can gain insight into the power of some of the analytic and statistical methods available. We shall classify engineering experiments in a number of ways, depending on the number of variables, whether important extraneous variables are present, the ways in which these variables interact, and so on. We shall not distinguish among industrial, production, research, development, pure, and applied experiments. Such classifications have, perhaps, meaning in sociological and political terminology, but they have little utility when subjected to scientific scrutiny and classification. The observation and measurement of northern auroras certainly qualify as a piece of pure research. Yet the data and their interpretation might look almost the same as the results of a study of truck traffic to and from Duluth over the past year, or an investigation of ceramic firing on days of differing humidity, temperature, and cooling rate.

Experiments and experimentalists differ, but virtually all follow a basic pattern, the same sequence of planning, operation, and analysis that is followed in the order of this book. Many experiments today, particularly those in the nuclear, electronic, and rocketry fields, are extremely expensive and at first glance very complicated. The reader who is acquainted with these fields may find it difficult to make the conceptual step from the immensely involved preparations, analysis, and precautions that precede a rocket flight or nuclear reactor startup to the quite basic and sometimes very simple experiments described in the following chapters. Actually, this is a case of the forest obscuring the individual trees. No matter how involved a test may appear to an observer, one can gain comfort from the knowledge that the reportable result will differ little in form (although, let us hope, it will differ much in quality) from the average college laboratory exercise reporting on an internal-combustion engine run, a test of a squirrel cage motor, or the calibration of a weir. A single rocket flight actually results in many reports—reports on motor performance, on down-range guidance, on flight-programming mechanisms, on cosmic-ray counts in outer space, on cloud cover as sensed by photocell, on the biological performance of test animals, and so forth. Each of these separate reports is handled by separate experts whose only common meeting ground is at the test vehicle itself.

Any test, no matter how complex in appearance, terminates in the transmission of data, conclusions, and perhaps suggestions to other persons. This information may be passed on by graph or curve, by mathematical equation or nomogram, by table, by statistical terms, or by the written word. Using a curve, we are limited to a result R versus variable X relationship, or an R versus X and Y function if parametric curves are used. Going to R versus X, Y, and Z requires several sheets or isometric coordinates. Beyond this level of functional complexity, we cannot graphically go, because the mind is incapable of visualizing relationships more complex than this. Shifting to equation form, we can represent the relation of R to more variables, but few experiments go beyond the simultaneous investigation of more than three independent variables at once.

Statistical presentations can be elegant and sophisticated, but what they tell us is summed up in a few words. A statistic can give information on a *population* of data and the variation of individual members within the population. It may give us information about the significance of a cause-and-effect relationship and tell us the probability of the occurrence of a certain event in the future based on what we have observed in the past. We might note here that we shall use statistical reasoning in two different parts of this book. First (in Chaps. 2 and 3), we shall see the statistic as a means of indicating the errors in instrument and measurement systems. Then (in Chaps. 8 and 9) statistics become a means of checking and/or revealing the significance of a test.

The transmission of data on a test by words, always a problem in the sciences, is the least efficient of all our reporting methods, yet one we can never ignore. There are undoubtedly certain tests being run today in physical laboratories with results that simply cannot be transmitted by the written word. In engineering, such tests are uncommon, and much of the usual engineering report is taken up with word descriptions and explanations.

Thus, an apparently complicated rocket or reactor test is really no more than a large number of discrete experiments tied together by an expensive collection of test equipment. Furthermore, although we may be able to imagine an extremely intricate test with results that could only be comprehended intuitively by hours of intensive study, it is doubtful that such a test would have much purpose. We might sense a mass of subtle interrelationships deep within the data, but unless these relationships are reducible to graph, equation, or words that can be assimilated by our professional colleagues, we have wasted our time. Most engineering tests should lead to action—a decision, more tests, or an admission of failure. Such results can occur only when we show others what we are doing.

In the author's opinion, many otherwise competent engineers carry out experiments that turn out to be expensive and poorly controlled because of a single conceptual failure. This is their total lack of self-questioning about the logic and reasoning behind each step they make. Too few engineers seriously ask themselves why a certain instrument is placed just where it is and why some other instrument is not used instead. This does not mean that one should ask trivial questions, such as "Why is that thermocouple attached there?" to which the answer is, quickly, "Why, to read the temperature, of course." Rather, we mean questions such as "Why are we using an iron–constantan couple soldered to the pipe elbow and wrapped with asbestos tape and connected to a 0-to-500-°F-range recorder with a print speed of 4 per minute and an accuracy (probable error? standard deviation?) of ±3%?" It is not likely that the average engineer will answer this series of questions in a hurry. We have raised queries, not just about a temperature measurement, but about the entire apparatus and the test itself. We have asked about the kind and attachment of the couple, anticipating a radiation error at the bond, perhaps progressive error increase as a result of corrosion or rust, perhaps even a problem of couple location that might make the time to reach steady state too long. We have asked about the range of the indicating recorder and about its error magnitude and thereby about the ranges and accuracies of all measurements, since they are intimately related through all experimental functions under study. We have asked about the print speed and therefore about the rate of data assimilation and the possibility of extraneous variations that may fluctuate within, say 15 s. In short, we have called many aspects of this experiment into scrutiny. Other questions might call for thought on the sequence and size of variable changes, on variable spacing over the apparatus range, on reruns where accuracy is doubtful, perhaps even on the choice of the variables themselves.

The reader may feel that tests get run regardless of whether such questions are asked, that measurement accuracy is always kept in mind in any test, that variables are changed until a good curve takes shape, that reruns are made when scatter is broad, and so on. This is certainly true. But when test engineers simply guess at the desired accuracy of each of a group of instruments, when they do not think at all about data-taking speed and extraneous variations, when they space the test runs by "feel" or intuition, when they ignore the possible biases introduced by a regular sequence of runs, and when they make reruns as a sort of desperate afterthought to "fix" a badly scattered curve, then the chances that the experiments will be long, costly, and inaccurate are excellent. The chances that the tests will reveal nothing useful at all are quite good. The chances that the engineers will happen upon the most efficient and well-controlled plans or that they will pick out some new and subtle effects are about nil. Far too much is made of such chance occurrences as the discovery of penicillin. Few important advances occur solely because of dirty culture bottles or sloppy test plans. It is the investigator who has methodically and completely thought through the possible extraneous effects and optimum methods of control who is most able to distinguish a truly unique and unusual effect from a host of side issues and extraneous-error causatives. Accidental discoveries occur when all foreseeable accidents have been computed, predicted, or eliminated, and only the new and novel accidents remain to happen.

Sometimes the most difficult job is asking the proper questions about an experimental plan. In this book we shall suggest what some of these basic questions should be and then proceed to methods of getting the answers.

1-2 DEFINITIONS AND TERMINOLOGY

Certain words and phrases will be used throughout the book in relation to the description of tests and test systems. Some are well established in technical English, whereas others are used by different authors in different ways. It is not practical to be too rigid about word use in a field as broad as experimentation, but the following meanings will apply in most of the work that follows.

We can break our testing "hardware" into three components: *instruments,* the *test apparatus* or rig, and the *test piece.*

Instruments sense, count, read, measure, observe, and then record, store, correct, read out what they have been set to "see."

The test apparatus is usually taken to be everything required to perform a test, including the instruments and the test piece.

The test piece is the specific hardware item that is undergoing test and may be replaced at will. Not every test apparatus has a test piece. For example, the test of a new production method using machines, workers, and material has no obvious test piece. If the test were designed to try out a new milling machine, however, then this machine would be the test piece.

The *experimental plan* is a general term relating to any specific set of test operating instructions that give sequence of test work, kinds and amounts of variable changes, and any replication decisions.

Sequence of testing refers to the order in which various operating changes are made in the test apparatus.

Replication means, in general, repetition, but more specifically, a return to a previous condition. For example, if we are testing a fan at a specific speed, pressure boost, and flow, we might take a series of repetitive readings without making any operating changes. This would not be replication. If we operate the fan at some new speed, boost, or flow condition and then go back to the previous condition and read again, we have replicated the first run. In agricultural and drug experiments, replication also implies a series of ground plots or animals that receive identical treatments.

Variable is used in a very general sense to imply any physical quantity that undergoes change. If we control this change independently of other quantities, we have an *independent* variable. If a physical quantity changes in response to the variation of one or more independent variables, we have a *dependent* variable. If a physical quantity that affects our test changes in a random and uncontrolled manner, we have an *extraneous* variable. We might have an electric motor in which input voltage and output load could be varied independently at the whim of the experimenter. The efficiency, winding temperature, and input current would be examples of dependent variables, and the surrounding temperature and line frequency variation (perhaps from a poorly stabilized generator) would be extraneous variables. Independent variables can usually be set at selected levels or values according to our experimental plan.

A *controlled* experiment is one in which the effects of extraneous variables have been eliminated and in which the independent variables can be varied exactly as the investigator chooses. *Control*, or the isolation of a test from its surroundings, is one of the great, basic conceptions of Western science.

Measurements are made of the independent, dependent, and often the extraneous variables by the instruments. We shall deal with a "best" value or "most probable" value for every measured quantity. This best value may always be made even better by installing more costly instruments, replicating the readings or test points, or hiring more skilled data takers. Thus, best is a relative term having definite utilitarian and economic overtones.

Measurements may be accurate or inaccurate depending on our standards of accuracy in a given test. In Chap. 2, we shall let the word *accuracy* refer to the fixed amount an instrument reading deviates from its known or *calibrated* input, regardless of how many times we make a measurement. Thus, an accuracy error is a *fixed* error.

Measurements may be precise or imprecise depending on how well an instrument can reproduce subsequent readings of an unchanged input. Thus, a *precision* error is not a fixed quantity like an accuracy error, but is different for each replication of a reading. Using statistical methods, we can assign single, *average* values to such precision errors in test instrumentation.

Error is a number—2 rpm, 0.6°F, 15 Ω, and so on—and is defined as the calibrated or known input reading minus the instrument reading. Error is thus known or predicted only when we can calibrate or otherwise check the test apparatus.

Uncertainty, like error, is a number, but an estimated one. S. J. Kline (personal communication) defined this uncertainty as "what we think the error would be if we could and did measure it by calibration." Uncertainty will always be analyzed in this text like a precision error, that is, through statistical methods.

To summarize these three kinds of deviation from the correct reading: We replicate a reading several times, and it deviates a fixed amount from the known input; we have accuracy error. We replicate a reading several times, and it deviates a different amount each time from the known input; we have precision error. We cannot calibrate or replicate (we may not even have the test rig built yet) any readings, but we can make an estimate of the error, which we think may be a fixed or varying type or some combination of the two; we have uncertainty.

To make a run, or *test run*, means simply to put the test apparatus in a certain fixed condition or *configuration* and record all the instrument measurements. Usually each test run will result, finally, in a *data point* or *test point,* where point implies an actual point on a real or imagined graph of the test results. In this book, point will be used in about the same sense as run, although usually in situations where an actual graphical point could result.

Data applies to all the "symbolic" products of an experiment; that is, data might be photographic images, magnetic impulses on tape, figures on sheets of paper, readings on mechanical counters, a simple yes-or-no answer in the mind of an observer, and so on. Data are not the pieces produced during a production test but are the written or remembered numbers of such pieces. *Raw data* are the symbolic materials obtained directly from the test instruments. *Processed data* are these symbolic materials after some additional mathematical operations have been performed on them, such as corrections by a calibration curve or plotting on a graph. Processed data that are plotted form a curve or may lead to some functional relationship among the independent and dependent variables that usually takes the form of a formula.

These relationships may be more or less significant in that a cause-and-effect relationship may be very obvious or not at all evident. Sometimes a simple engineering curve will be sufficient to reveal such relationships, whereas at other

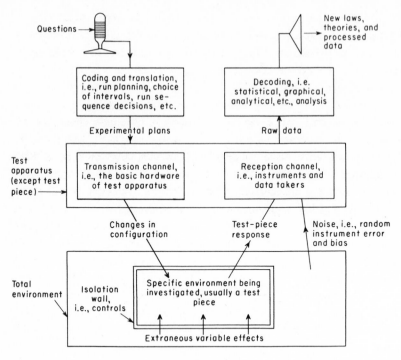

Figure 1-1 A simplified block diagram of a typical experiment viewed as a communications network. In the sense of this diagram, the major aims of this book are to (1) ensure good "coding" and effective planning, (2) strengthen the "wall" blocking our specific environment from the outside, (3) reduce the "noise" due to instrument fluctuation, and (4) "decode" in such a way as to maximize the useful information coming from our fictive "loudspeaker."

times we may choose a statistical *test of significance*. Statistical analysis is intimately related to distributions of test readings or other data. When discussing the general theories of statistics, we shall often deal with *populations* of errors, data, and so on, which this book will take as the total set of an infinity of such errors, data, and so on. Thus a *parent distribution* is simply a population of whatever we are discussing. From this infinity of readings, we obtain through our test a *sample* of readings, finite in number. The larger the sample, the more closely (we hope) its distribution will approximate the distribution of the population. Figure 1-1 shows some of these ideas in schematic form.

1-3 NOMENCLATURE

A great many formulas from diverse technical fields will be used in both the text of this book and the problems at the ends of the chapters. In general, the exact technical meaning of the symbols is not important, for we are stressing here general experimental methods rather than specific applications of engineering theory. In most cases, enough engineering explanation is given so that readers can orient themselves in the specific technical area. All of the in-text examples are based on specific, known experiments.

The beginning of the alphabet in both upper- and lowercase (A, B, C, ... and a, b, c, ...) is used for a variety of purposes but most commonly to represent constants in experimental equations. The end letters in uppercase (W, X, Y, Z) represent variables, usually controlled, or simple coordinates on graphs. End letters in lower case (w, x, y, z) represent deviations of actual or measured values from the correct or calibrated value, which can be identified by the subscript c. Thus $X_c - X = x$, and so on.

In the statistical sections, we shall use Greek letters to refer to quantities used in the evaluation of *population* parameters and Latin letters for formulas involving *sample* parameters. Table 1-1 gives some of the special symbols that are commonly used.

Table 1-1 Nomenclature

Symbol	Definition
A, B, C, ...	Equation constants or special symbols
a, b, c, ...	Equation constants or special symbols
E_n	Expected number of occurrences (Sec. 9-3)
f	Function of
h	Index of precision for a sample (Chap. 2)
ln	Log to the base e
N	Dimensionless number (usually with subscript)
n	Number of items (such as groups, readings)
P	Probability of an event (defined in Chap. 2)
p	Probable error (Eq. 2-24)
R	Result or dependent variable
s or SD	Standard deviation for a sample (Eq. 2-21)
s_m	Standard deviation of the mean (Eq. 2-25)
T	Dimensionless deviation for a sample (Eq. 2-22)
w	General reference deviation (Chap. 3)
W, X, Y, Z	Independent or dependent variables
$\bar{W}$, $\bar{X}$, $\bar{Y}$, $\bar{Z}$	Mean or average value of independent variable for a sample
W_c, X_c, Y_c, Z_c	Best or correct value (which may be the mean value)
Δ	Small change or increment
η	Index of precision for a population (Sec. 2-4)
μ	Best value or calibrated input for a population (Chap. 2)
σ	Standard deviation for a population (Eq. 2-6)
τ	Dimensionless deviation for a population (Eq. 2-9)
ϕ	Probable error for a population (Chap. 2) or meaning "function of"

1-4 THE INTERNATIONAL SYSTEM OF UNITS

Engineering in the United States and the United Kingdom is now facing a substantial shift in unit systems, thread sizes, and other dimensional constants as the world standardizes on "Le Système International d'Unités", or the so-called SI, or metric, system of units. Actually, physics and chemistry in the United States have long used metric systems, usually either the centimeter-gram-second (cgs) system or the meter-kilogram-second (mks) system. The present SI system has grown out of the mks system following a 1960 meeting of representatives from 40 countries, and conversion to this system is now taking place worldwide. In the United States, it is evident that many years will pass before conversion is complete, and texts during this period should work in both the English and SI systems so as to aid and hasten the transition. Appendix A of this text gives selected conversions of English units into SI units for those wishing to work all book problems and examples in the SI unit system.

Base units in SI are length in meters (m), mass in kilograms (kg), time in seconds (s), electric current in amperes (A), thermodynamic temperature in kelvins (K), amount of matter in moles (mol) and luminous intensity in candelas (cd). A huge variety of other units and constants also have standardized names and abbreviations, following the general rule that the unit when written out is *not capitalized* but when abbreviated *is capitalized if it is a proper name.* Examples are newton (N), pascal (Pa), watt (W), and so on. In this book, since some SI units may be unfamiliar to many starting engineers, the full name is generally used, at least the first time it appears in an example.

The central problem with unit systems for most newcomers to engineering is the way gravity and Newton's second law are handled. In the English engineering system, two methods have been in common use for years. If we write this law as

$$F = Ma \tag{1-1}$$

a might be in feet per square second (ft/s^2), M in slugs, and F then is in pounds of force. If we write it as

$$F = M \frac{a}{g_c} \tag{1-2}$$

then M might be in pounds-mass, a in feet per square second, g_c a fixed and universal constant having the value 32.2 lb$_m \cdot$ ft/lb$_f \cdot$ s^2 so that F is again in pounds-force. The only difference between these two approaches is the replacement of the slug by the pound-mass through the use of g_c.

In the SI system, Eq. 1-1 is used with a in meters per second, M in kilograms, and F is then in newtons, the basic SI unit of force. Since the acceleration of standard gravity in SI is 9.8 m/s^2, the gravitational force in newtons exerted by a mass of M kg can be found from Eq. 1-1 by letting a be 9.8 m/s^2. It is evident, then, that the newton is a derived unit having basic SI units of kg $\cdot$ m/s^2.

Similarly, energy is expressed in joules (J), which are newton-meters; power in watts (W), which are joules per second; and stress or pressure in pascals (Pa) or newtons per square meter.

PROBLEMS

1-1 A 5-lb mass exerts what force on a spring scale calibrated in newtons at standard gravitational acceleration?

1-2 The angular acceleration I is given by $\frac{1}{2}Mr^2$. If M is 10 lb$_m$ and r is 0.5 ft, find I in kilograms-meter square.

1-3 If the specific heat of air is 0.24 Btu/lb$_m$ $\cdot$ °F find its value in joules per kelvin.

1-4 What are the SI base units for power (joules per second) and for pressure or stress (pascals)?

1-5 If the gravitational attraction is reduced by a factor of 6, what will 1 kg show in a spring scale calibrated in newtons?

1-6 What is the pressure in pascals at the deepest point in the ocean (36,000 ft deep)? Take the density of water as 64 lb$_m$/ft^3.

1-7 A 20-lb$_m$ weight is moving 8 ft/s. What is its kinetic energy in joules?

1-8 A solar panel receives 220 Btu/hr $\cdot$ ft^2 at noon and converts 10% of it to electricity. If the area is 50 ft^2, what is the power generated in watts?

BIBLIOGRAPHY

American Society of Mechanical Engineers: *SI Series Booklets,* SI-1 through SI-11; ASME, New York, 1976.

Baird, D. C.: *Experimentation: An Introduction to Measurement Theory and Experimental Design,* Prentice-Hall, Englewood Cliffs, N.J., 1962.

Beverage, W. I.: *The Art of Scientific Investigation,* W. W. Norton, New York, 1950.

Cohen, M., and E. Nagel: *An Introduction to Logic and Scientific Method,* Harcourt, Brace & World, New York, 1934.

Cook, B. H., and E. Rabinowicz: *Physical Measurement and Analysis,* Addison-Wesley, Reading, Mass., 1963.

Holman, J.: *Experimental Methods for Engineers,* McGraw-Hill, New York, 1966.

Sci. Am., vol. 199, no. 3, September 1958. Entire issue devoted to "Innovation in Science."

Wightman, W. P.: *The Growth of Scientific Ideas,* Yale University Press, New Haven, Conn., 1953.

Wilson, E. B.: *An Introduction to Scientific Research,* McGraw-Hill, New York, 1952.

TWO

THE STATISTICS
OF INSTRUMENT ERROR

The subject of statistics is the *mathematical properties of distributions of numbers.* Any group of numbers of any size (including a single number) can be examined using statistical tools, and two or more such groups can be compared using various statistical methods. These distributions of numbers can be generated in a great variety of ways. They may flow from natural phenomena (such as a compilation of daily temperatures or cosmic-ray counts), from human activity (such as wrong numbers per day in a telephone exchange or the annual gross national product of the United States), or from the minds of statisticians. Through the years, many mathematical statisticians have proposed certain distributions of numbers as typical or representative of "natural" distributions. These imaginary or theoretical distributions are necessary to the science of statistics because of their simplicity. Manipulating a group of, say, 50 numbers is much easier if we can represent those 50 numbers by two or three basic numbers that adequately characterize the complete distribution.

In the usual statistics course, these ideal distributions are studied in their mathematical forms and the application of the subject to natural distributions is treated after these idealizations are completely revealed. In this book, however, we shall approach the subject of statistics in a reverse manner: through the study of that particular set of natural groups of numbers produced by technical and scientific instruments as they supply us with information about the vast collection of phenomena and devices in which humanity has interest. In short, the bulk of this chapter (Secs. 2-1 through 2-9) will introduce the specialized yet essential subject of *instrument statistics,* that part of statistical theory that has been found to be useful in dealing with instruments and test apparatus of every sort.

Once we have performed this wedding of the problem of instrument variability to the mathematical study of number distributions, we shall gain a tool that has other uses in the general field of experimentation. In particular, we shall obtain the characteristics of the *normal,* or *Gaussian, distribution* as it relates to instrument error, then show (in Sec. 2-10) how this useful statistical model can be utilized to study the *results* and interpretations of experimental tests as well as the deficiencies of their instruments. The overall scheme is to organize the analytic tools available to experimenters in more or less the same order they would meet them, first in planning, then running, and finally analyzing an engineering test. Section 2-10 is placed ahead of where it would normally appear in following this scheme, but this layout will allow us to examine these simple but crucial statistical ideas from several practical directions, one after the other.

Consideration of measurement error is the first essential step in any test program. If we cannot obtain measurements of adequate accuracy to solve the problem, there is no point in planning any other aspect of the experiment. But we must do more than simply define the error characteristics of the several instruments of our test. Instrument readings are often combined in some mathematical way to obtain a final result. For example, the efficiency of a machine is usually investigated by measuring the input energy and the output energy and comparing the two. Both may be uncertain, because of measurement error, and both will contribute to an uncertainty in the efficiency. This combination of errors, called *error propagation*, will be studied in Chap. 3, after we have seen here how individual instrument variability can be analyzed statistically.

2-1 KINDS OF ERROR

We shall postulate three basic error sources in physical measurements. These are as follows:

1. Failure of the primary sensing element to reflect the measured quantity; for example, a thermocouple junction that is corroded or loose, or that has radiation and conduction losses that make the junction temperature different from the surrounding temperature.
2. Failure of the indicating or secondary portion of the instrument to reflect faithfully the sensing-element response; for example, a potentiometer that gives an incorrect millivolt reading when fed by a thermocouple, because of improper standardization, calibration, or malfunction of its mechanical or electrical components.

3. Failure of the observer to record the instrument reading correctly; for example, a person who reads the wrong scale on the potentiometer dial.

Although all three sources of error may be present in a given determination, we expect that one or another may be the major trouble factor. The skilled investigator will quickly see the weak link in this three-link chain of information transmission. We have said nothing at this point of the fourth, and perhaps most critical, link, namely, the interpretive skill of the experimenter in gaining the maximum truths from the collected data. This final link can be discussed only after the data have been gathered, and we shall consider it, as best we can, in later chapters.

These three sources produce two basic classes of error in an experimental determination. The classes are *errors of precision* and *errors of accuracy*, and we expect all readings to possess an overall error compounded of both classes but in different relative amounts. The relative weight of either depends on the instrument or test situation.

Precision error is always present when successive measurements of an unchanged quantity yield different numerical values. For example, a tachometer measures the speed of a constant-speed motor, obtaining in successive measurements 1050, 950, 1000, 1030, 990, and 980 rpm. If the motor is absolutely stabilized* on a given speed, we conclude that the tachometer gives data that are not perfectly precise.

Accuracy error is always present when the numerical average of successive readings deviates from the known correct reading and continues to deviate no matter how many successive readings are made. For example, in the case of the tachometer, the average of the six readings is 1000 rpm. If this checks a standard or calibration reading of 1000 rpm, we may infer that little or no accuracy error exists. If the standard value is known to be 950 rpm, we infer that the tachometer is *not* accurate and at this speed reads about 50 rpm high.

In this chapter, accuracy and accuracy error refer to samples or groups of readings sufficient in number to establish whether the instrument is really scattering around the "true" value or whether it shows a continuous bias or "one-sided" deviation. Practically, we would replace or repair a tachometer that deviated by 50 rpm out of 1000, whereas we might prefer to simply calibrate and use the one that always reads 50 rpm high.

---

*In this chapter we shall assume that the measured quantity is not changing and that all variation is due to lack of precision of measurement. Should the variation actually be due to poor control over the variable, all the analysis we undertake will still hold, but the "fix" will be different. In many tests, the experimenter will sense which is the case.

The establishment of which of these two kinds of error we have requires that a calibration or similar test be made. Perhaps the instrument manufacturer will give us such data, or perhaps we can obtain them ourselves. In the planning stages, however, we may not know anything about the kinds of error beyond anticipating that some deviations from the correct value must be expected. In such cases we deal with uncertainty rather than error (as noted in Chap. 1), and we infer, compute, or (more likely) guess the magnitude of this uncertainty. In this book, such uncertainty, after it has been estimated, will be treated exactly the same as precision error. Thus everything we say about precision error will apply to uncertainty, even though our uncertainty may be (and probably is) made up of both precision and accuracy components.

2-2 CALIBRATION

We calibrate an instrument by presenting it with one or more known inputs and noting its output readings. By comparing the input and output graphically or in a table, we provide the user of the instrument with a *calibration curve,* a means of correcting the actual reading to the "correct" reading based on the standard inputs. Now, *calibration can correct only for accuracy error.* If we provide a single known and fixed input and the instrument output is different each time we read it, we have no way of knowing in which direction and of what size the error will be with any given reading.

Calibration is usually the first step in any test involving instruments, although in many cases we may accept the manufacturer's calibration, that is, the scale in the instrument that the manufacturer provides. A calibration experiment will often tell quite a bit about an instrument, its relative accuracy, and its precision problems.

Example 2-1 The accompanying table shows results of a calibration test of a carbon dioxide meter on two different days. Three tanks, each of known CO_2 percentage, were available, and the student made a single reading from each tank on each of the two days.

Calibration table–CO_2 meter

Tank (standard) percent CO_2	1.22	2.83	4.07
Percent CO_2 from meter, Oct. 13	1.55	3.55	4.91
Error, Oct. 13 (percent inst. − percent tank)	+.33	+.72	+.84
Percent CO_2 from meter, Oct. 25	1.45	3.08	4.60
Error, Oct. 25 (percent inst. − percent tank)	+.23	+.25	+.53

This type of meter, incidentally, absorbs CO_2 in aqueous solution similar to the Orsat, combustion gas analyser, but is intended for use where humans are confined in badly ventilated spaces. How should we plot these data, what do they reveal, and what more, if anything, should be done?

Solution There are three common ways to plot calibration data: (1) output versus input; (2) difference between output and input versus input; and (3) percent error in output versus input. Figure 2-1 shows the data plotted as output versus input; this graphical presentation is often called a calibration graph or a *parity graph* because it allows comparison of two sets of numbers that, if all were well, would be equal (have parity). A parity graph is plotted with identical scales on the X and Y axes. A straight line thereon, the *parity line*, drawn through the origin at 45°, allows us to rapidly check the deviations of the instrument output from the known input. Figure 2-1 reveals the following: (1) The instrument reads high (all points are above the parity line). (2) The instrument gives reasonable linear and consistent results on each day but reads differently on the two days. (3) The error on each day appears to be roughly proportional to the CO_2 input concentration, but there is really not enough data to prove this.

Clearly, we do not yet have enough information on this CO_2 analyzer. The student should take a series of identical readings on a single day and see

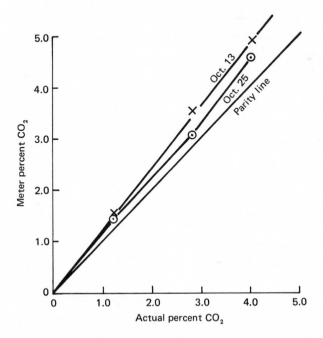

Figure 2-1 Parity calibration graph for Example 2-1 with each day plotted separately.

if the instrument gives the same readings over short time periods. If it does then it should be calibrated before each use. Probably, the instrument has some sensitivity to surrounding temperature and possibly other factors of storage and use.

Assuming that the three points found for October 13 are correct for that day, we then form the *calibration curve* for October 13 by connecting the three points (and the origin) by straight lines, *not* by drawing the best straight line through the four points. The reason we do not do this is that we have not established that the instrument reads in a completely linear manner, that is, that it always gives a straight-line response to changes in CO_2. Whenever we lack knowledge of the functional relationship between the true and gauge readings, it is always best simply to connect the calibration points by straight lines.

We shall see the parity graph again, for it has many uses in experimental analysis. When the data does not have the relatively large errors of Fig. 2-1, however, other methods of presentation may show more.

Example 2-2 The accompanying table gives the result of a calibration test on two scuba depth gauges, gauge A being a *bourdon* type having a hollow spring moving in a pressure-resistant case and gauge B being a so-called capillary gauge in which water pressure forces a trapped bubble of air into a scaled tube. The test was carried out in a small, transparent pressure chamber filled with water pressurized with a hand pump, with its interior pressure read on a calibrated pressure gauge.

Calibration of two depth gauges

True depth (ft)	0	10	20	30	40	50	60	70	80	90	100
Bourdon											
descent (ft)	0	13	23.9	34.5	44.5	54.0	63.4	73.6	82.6	92.8	101.8
ascent (ft)	3	14.5	24.5	34.8	44.8	54.6	64.3	73.9	83	93.5	101.8
Capillary											
descent (ft)	1.5	10.4	19.9	30.4	39.5	48.5	57.8	66.8	75.5	85.0	93.0
ascent (ft)	1.8	11.5	21.0	30.5	39.8	49.0	58.5	67.3	76.3	85.8	93.3

Again, we wish to plot the data and to consider the accuracy of these two pressure-measuring instruments.

Solution The deviations shown in the table are quite small and may not reveal their character on a parity graph. Figure 2-2 shows another presenta-

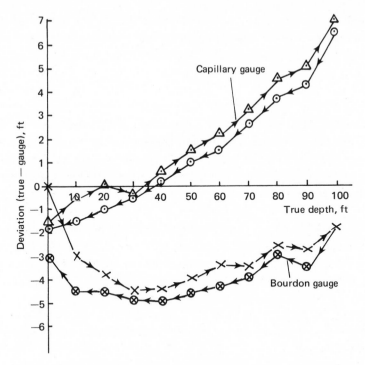

Figure 2-2 Deviation graph showing the hysteresis of two scuba depth gauges.

tion that we might call a *deviation plot.* We obtain it by plotting the difference between the true, or chamber, depth and the instrument-predicted depth, against true depth. A positive point is a reading that was incorrectly shallow. Note that Fig. 2-2 is *not* a calibration curve, since we cannot read the instrument depth into the graph and then correct it, as was possible with Fig. 2-1.

By distinguishing between the points taken during the descent (pressurization) from those taken during ascent (pressure decrease), we immediately establish that both gauges have *hysteresis*; that is, they will give a different reading at a given depth if that depth has been approached from a greater rather than a lesser depth. In the case of these two gauges, Fig. 2-2 shows that they read between $\frac{1}{2}$ and 1 ft lower on the descent than on the ascent. Hysteresis is a characteristic of pressure gauges and many other instruments.

Gauge A is a practical example of an instrument with a relatively constant offset error, what we shall call a *sum-type error.* If we subtract about 3.5 ft from all of gauge A's readings, we shall be within a foot of true depth except at the extreme ends of the scale.

Gauge B, on the other hand, illustrates another common type of instrument accuracy error, a *product-type error*. Notice that the gauge shows a continuous and almost linear deviation from 2 ft too *deep* at the surface to 7 ft too *shallow* at 100 ft. If we multiply the true depth by the factor, 0.92, and add 2 ft, we obtain an approximation of the Fig. 2-2 curves for gauge B. Evidently, the manufacturer offset the scale 2 ft to locate zero error at about 30 ft of true depth. This gives the gauge ±2 ft accuracy in the range 0 to 60 ft, the depth range of most amateur scuba diving. If, however, this gauge were trusted when it read 93 ft with the diver actually at 100 ft, the diver might remain too long and, if he or she surfaced immediately, get a case of decompression sickness.

Offset, or sum-type errors, often come about from slipped scales, bent meter pins, or incorrectly estimated tare weights on scales. *Product-type errors* occur whenever the error is in some way proportional to the reading, as might occur with a weak magnet in a meter movement or a spring scale with a weak (or too strong) spring. Calibration often reveals these characteristic patterns.

2-3 THE HISTOGRAM

Although the instruments we calibrated in the previous section were treated as though they had no precision error, the actual data indicated otherwise. The CO_2 meter is clearly affected by some day-to-day effect (Fig. 2-1), whereas both the depth gauges gave different readings depending on the way the reading was approached (Fig. 2-2). Clearly, all these instruments in normal use, where calibration may not be possible, would show some precision error over a period of activity; that is, they would, when presented with a given input, give different outputs depending on external factors having nothing to do with the measured quantity.

It is obviously essential that any proper calibration procedure detect the presence of such random errors. One way to do this is to provide the instrument with the same input signal time after time and record the instrument's output for each trial. A fuse might be regarded as a simple instrument that switches at a specified electrical current flow. Table 2-1 gives the results of an experiment on

Table 2-1 Failure current of $\frac{1}{16}$-A miniature fuses

Failure current, I (mA)	92	93	94	95	96	97	98	99	100	101
Number of fuses ($n = 69$)	3	4	9	10	13	10	8	8	2	2

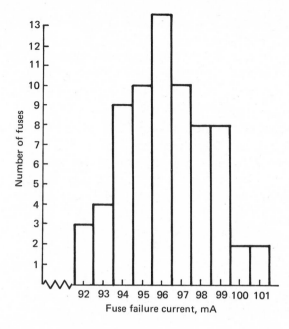

Figure 2-3 Histogram of 69 fuse-failure currents as given in Table 2-1.

$\frac{1}{16}$-A-rated miniature fuses from a single manufacturer. The fuses were subjected to a linearly increasing current with a fixed rate of increase and the current at which each fuse blew was recorded to the nearest milliampere.

The data in Table 2-1 can be displayed in an equally compact and more revealing form on a *histogram*. We group the data by dividing the data range into convenient, equally sized sequents. In this case, the obvious division is 1 mA; since this is the least division of the data in the range of 92 to 101 mA. A good general rule when dealing with numbers of data between 10 and 100 is to form the histogram by dividing the data range into from 5 to 10 segments. Here, the 1-mA division provides 10 segments and gives a good picture of the distribution, as Fig. 2-3 shows. Of course, if you have more data, you can break your data base into smaller segments. Figure 2-13 is an example of this.

The histogram shows us how the data groups together. We note, for example, that the fuse-failure data has a *central tendency*, that is, a tendency to group around a central value (about 96 mA in this case). The distribution of numbers has *tails*; that is, the number of fuses failing at progressively higher and lower currents grows less and less as we go farther from the central value.

As a calibration experiment, Fig. 2-3 shows us that these fuses fail at a considerably higher current than their rated $\frac{1}{16}$ A (67.5 mA). Thus, if this rate of current overload is typical of the service overloads expected for the fuses, we should relabel them as "$\frac{1}{10}$-A" fuses, since the great majority of them fail at just under this value.

There is, however, a great deal more to be learned from Fig. 2-3. This bell-shaped curve is an experimental approximation of the *normal curve of error*, a theoretical curve with simple properties that approximates the many calibration and instrument curves having the form of Fig. 2-3. By using this theoretical and mathematical model, we can answer a number of questions about distributions of data of this sort. For example, we can predict how many fuses we would need to test to find one that failed at, say, 91 mA, or we can predict what proportion of fuses will fail between, say, 98 and 100 mA. In addition, we can subject data like Fig. 2-3 to tests that will suggest how well the mathematical model matches this particular sample of data.

2-4 THE NORMAL CURVE OF ERROR

When only accuracy error exists, its mathematical correction through calibration curves, like Fig. 2-1, is a simple matter. Precision error is more complex and gives us data distributions like Fig. 2-3. If we could continue to test fuses and use a test apparatus that would read the failure current more closely, say, to within a hundredth or even a thousandth of a milliampere, then we would build up a histogram with many more segments than in Fig. 2-3, finally approaching a smooth curve similar to that shown in Fig. 2-4. This smooth *distribution curve,* or *density function,* then represents the *infinite,* or *parent, population* of fuse data from which we drew a sample of 69 fuses to obtain Fig. 2-3, the histogram.

Figure 2-4 is a sketch showing the form and nomenclature of the famous *normal curve of error,* or *Gaussian distribution.* The equation of this distribution,

$$y = y_0 \exp\left(-\eta^2 x^2\right) \tag{2-1}$$

is often derived[*] from the following two basic assumptions:

1. The final error in any observation is the result of combining a large number of very small errors, all of equal magnitude.
2. These small errors have equal probability of being positive or negative.

Now, in fact, if we consider the various components of a simple fuse and how their various tolerances and variations might affect the blow-out current,

---

[*]For example, see Worthing and Geffner, 1943, pp. 148–153; or Young, 1962, pp. 151–157.

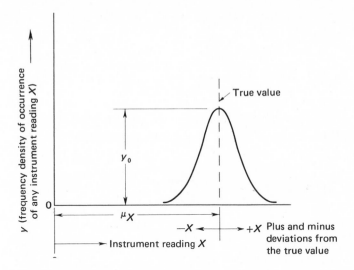

Figure 2-4 A sketch of the frequency density of any reading from a normally deviating instrument as a function of its deviation from the population true value.

and further, how the aging of the fuse and its condition of storage might affect these components, then it is apparent that the two assumptions just noted do not represent the real situation of the fuses or, for that matter, any other instrument.

The reason that Eq. 2-1 is the most important mathematical model of instrument error is simply that it fits more real cases of instrument and experimental variability than any other, a point we hope to establish in this and later chapters.

Professional statisticians are very specific about their nomenclature; when they are dealing with a *parent population* or *theoretical distribution,* such as specified by Eq. 2-1, constants are given Greek letter symbols. When dealing with a sample drawn from a parent distribution, such as Fig. 2-3, then Roman letters are used for constants. Thus in Eq. 2-1, the constant η is the *modulus* or *index of precision,* y is the frequency density of occurrence at any deviation x away from the central point μ_x, and y_0 is the frequency density of occurrence at this central point. Holding y_0 and η constant and plotting y versus x gives the symmetrical, bell-shaped curve of Fig. 2-4. This curve encloses the entire population of deviations from a given instrument, and we are interested, first, in the mathematical expression for the area A under the curve. Using the calculus, this is found from

$$A = 2 \int_0^\infty y_0 \exp\left(-\eta^2 x^2\right) dx \qquad (2\text{-}2)$$

Equation 2-2 is a definite integral of some complexity. It may be evaluated[*] or may be found in most tables of integrals.[†] The area is

$$A = \frac{\sqrt{\pi}}{\eta} y_0 \tag{2-3}$$

It is convenient to let this area take the value of *unity*. Then $y_0\sqrt{\pi}/\eta = 1$, and $y_0 = \eta/\sqrt{\pi}$. Equation 2-1 becomes, as a result of this *normalizing procedure*,

$$y = \frac{\eta}{\sqrt{\pi}} \exp\left(-\eta^2 x^2\right) \tag{2-4}$$

and y will then have the units of η, which in turn must have the reciprocal units of x.

Now, y is not a particularly useful quantity in its own right. In almost all situations, what we wish to know is the *probability* that any given reading will occur. Since we have made the total area under the curve equal to unity (or 100%), any lesser area within this total must represent that fractional chance that a given reading within the lesser area will occur. The probability that we shall obtain a reading *within* the deviations $\pm x$ on Fig. 2-5 is equal to the area

[*]Worthing and Geffner, 1943, pp. 155–157.
[†]Such as Burington, 1949.

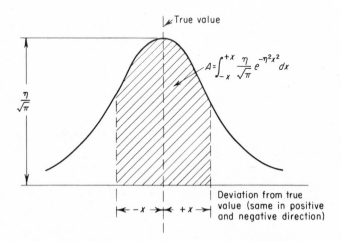

Figure 2-5 A sketch of the normalized density function for the normal distribution using the nomenclature of Eq. 2-5.

between these limits, shown shaded in the figure. The chance that we obtain a reading *between* the central point and the $+x$ limit is half as great. The chance that a reading will occur *outside* the plus and minus limits of x is equal to unity minus the shaded area.

Mathematically, we can express the shaded area between $+x$ and $-x$ as

$$P_x = \int_{-x}^{+x} \frac{\eta}{\sqrt{\pi}} \exp\left(-\eta^2 x^2\right) dx \tag{2-5}$$

2-5 PRECISION ERROR
AND STANDARD DEVIATION

The modulus of precision η is not an especially handy parameter to describe the spread of a population, since it is not readily computed from samples. Instead, statisticians have discovered a *precision index* (a quantity that measures the *spread* or *precision* of a distribution), which allows the computation of probabilities directly from sample data.

This quantity, the *standard deviation* σ, (or *variance*, which is σ^2) is defined as that deviation value equal to the square root of the sum of the squares of all the deviations, divided by the total number of such deviations n.

$$\sigma = \left(\frac{\sum_{n=1}^{n=\infty} x^2}{n}\right)^{1/2} \tag{2-6}$$

But

$$\sum_{n=1}^{n=\infty} x^2 = \int_{-\infty}^{+\infty} x^2 y \, dx \quad \text{and} \quad n = \int_{-\infty}^{+\infty} y \, dx = 1.0$$

for the normalized expression of the normal distribution. Then Eq. 2-4 inserted in Eq. 2-6 with the above gives

$$\sigma = \left(\frac{\eta}{\sqrt{\pi}}\right)^{1/2} \left[\int_{-\infty}^{+\infty} x^2 \exp\left(-\eta^2 x^2\right) dx\right]^{1/2} \tag{2-7}$$

which is best handled by recourse again to a table of definite integrals. The result is

$$\sigma = \frac{1}{\eta\sqrt{2}} \tag{2-8}$$

Let us now define a *dimensionless deviation* τ as the ratio of the deviation x from the central value μ_x to σ, so that

$$\tau = \frac{x}{\sigma} \tag{2-9}$$

These two equations give the relationship between η and τ,

$$\eta = \frac{\tau}{x\sqrt{2}} \tag{2-10}$$

and Eq. 2-5 becomes

$$P_x = \frac{1}{\sqrt{2\pi}} \int_{-\tau\sigma}^{+\tau\sigma} \frac{1}{\sigma} \left(\exp \frac{-\tau^2}{2} \right) dx$$

This can be rewritten more conveniently as

$$P_\tau = \frac{1}{\sqrt{2\pi}} \int_{-\tau}^{+\tau} \exp \frac{-\tau^2}{2} \, d(\tau) \tag{2-11}$$

Equation 2-11 is a compact form of the *probability integral*, which is usually evaluated from a tabulation such as Table 2-2. Table 2-2 and Eq. 2-11 give values of P_τ for the so-called two-tailed probability, that is, the chance of finding a reading between $+(x/\sigma)$ and $-(x/\sigma)$. As x and τ approach infinity, P_τ approaches 1.0. Thus for the *two-tailed* case,

$$0.0 \leqslant P_\tau \leqslant 1.0 \tag{2-12}$$

The *one-tailed* probability gives the chance of finding a deviation x between the center of the distribution μ_x and *either* $+\tau$ or $-\tau$. Then

$$0.0 \leqslant P_\tau \leqslant 0.50 \tag{2-13}$$

Table 2-2 Two-tailed probability tables for the normal integral[a]

τ	P_τ	τ	P_τ	τ	P_τ	τ	P_τ
0.0	0.0000	0.5	0.3829	1.0	0.6827	1.5	0.8664
0.02	0.0160	0.52	0.3969	1.02	0.6923	1.55	0.8789
0.04	0.0319	0.54	0.4090	1.04	0.7017	1.6	0.8904
0.06	0.0478	0.56	0.4245	1.06	0.7109	1.65	0.9011
0.08	0.0638	0.58	0.4381	1.08	0.7199	1.7	0.9109
0.1	0.0797	0.6	0.4515	1.1	0.7287	1.75	0.9200
0.12	0.0955	0.62	0.4647	1.12	0.7374	1.8	0.9281
0.14	0.1113	0.64	0.4778	1.14	0.7457	1.85	0.9357
0.16	0.1271	0.66	0.4907	1.16	0.7540	1.9	0.9426
0.18	0.1428	0.68	0.5035	1.18	0.7620	1.95	0.9488
0.2	0.1585	0.7	0.5161	1.2	0.7699	2.0	0.9545
0.22	0.1741	0.72	0.5285	1.22	0.7775	2.1	0.9643
0.24	0.1900	0.74	0.5407	1.24	0.7850	2.2	0.9722
0.26	0.2051	0.76	0.5527	1.26	0.7923	2.3	0.9786
0.28	0.2205	0.78	0.5646	1.28	0.7995	2.4	0.9836
0.3	0.2358	0.8	0.5763	1.3	0.8064	2.5	0.9876
0.32	0.2510	0.82	0.5878	1.32	0.8132	2.6	0.9907
0.34	0.2661	0.84	0.5991	1.34	0.8198	2.7	0.9931
0.36	0.2812	0.86	0.6102	1.36	0.8262	2.8	0.9949
0.38	0.2961	0.88	0.6211	1.38	0.8324	2.9	0.9963
0.4	0.3111	0.9	0.6319	1.4	0.8385	3.0	0.9973
0.42	0.3255	0.92	0.6424	1.42	0.8444	3.5	0.999535
0.44	0.3401	0.94	0.6528	1.44	0.8501	4.0	0.999937
0.46	0.3545	0.96	0.6629	1.46	0.8557		
0.48	0.3688	0.98	0.6729	1.48	0.8611		

[a]P_τ defined in Eq. 2-11; τ defined in Eq. 2-9.

and the one-tailed probability is simply half the two-tailed value. Notice that a deviation equal to ± 1 standard deviation σ ($\tau = 1.0$) encloses about 68.3%, or about two-thirds, of all the population. A deviation of a 2σ ($\tau = 2.0$) encloses about 95.5% of all the population, and this is often taken as approximately 95%, or *20-to-1 odds*. That is, the chance of finding a reading outside $\pm 2\sigma$ is about 1 in 20, a common probability chosen to include "most" or "almost all" the population.

2-6 THE BEST READING FROM SAMPLES

The normalized form of the normal law, Eq. 2-4, eliminates the y_0 term in Eq. 2-1, but we still must determine the "true value" μ_x before we can find any x values, since

$$x = X - \mu_x \qquad (2\text{-}14)$$

Our fuses in Table 2-1 were rated at 67.5 mA, but this can hardly be taken as a measure of the center of the Fig. 2-3 distribution. We might thus pose two questions: What is the best estimate of the true value for the fuse blowout data? How is this estimate affected by the number of readings we take?

Consider a sample of n readings having values X_1, X_2, X_3, ..., X_n that result from repeated measurements of an unchanged quantity. We shall assume that these readings are part of a normal population with a correct value X_c that is unknown. The reading X_1 falls within a tiny interval Δx, as shown in Fig. 2-6. Let us write Eq. 2-4 in terms of Eq. 2-8, obtaining

$$y = \frac{1}{\sigma\sqrt{2\pi}} \exp\left(-\frac{x^2}{2\sigma^2}\right) \tag{2-15}$$

Now the chance ΔP_1 of getting the reading X_1 is the area of the little rectangle $y \, \Delta x$ or, using Eq. 2-15,

$$\Delta P_1 = \frac{\Delta x}{\sigma\sqrt{2\pi}} \exp\left[-\frac{(X_c - X_1)^2}{2\sigma^2}\right] \tag{2-16}$$

Similar expressions will hold for the probability of occurrence of X_2, X_3, and so on. The probability that the entire sample of n readings will occur is equal to the product of the probabilities that each single reading will occur, or $\Delta P_1 \times \Delta P_2 \times \Delta P_3 \times \ldots \times \Delta P_n$. This basic assumption from probability theory

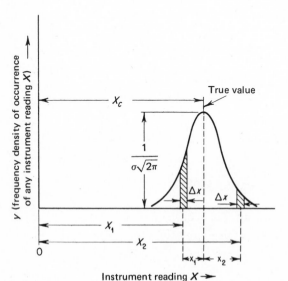

Instrument reading X ➡

Figure 2-6 A sketch showing the nomenclature of Eq. 2-15.

can be demonstrated by coin tossing. The chance of a single head is 50% or probability 0.5. The probability of two consecutive heads is 0.5×0.5, or 0.25; and of three consecutive heads, $0.5 \times 0.5 \times 0.5$, or 0.125; and so on. Then if ΔP_{total} is the overall probability that the set of n readings will occur,

$$\Delta P_{total} = \left(\frac{\Delta x}{\sigma \sqrt{2\pi}}\right)^n \exp\left[-\frac{(X_c - X_1)^2 + (X_c - X_2)^2 + \ldots + (X_c - X_n)^2}{2\sigma^2}\right]$$

$$(2\text{-}17)$$

we shall assume that the collective set X_1 through X_n will be related to X_c such that the total probability ΔP_{total} will be a maximum. This leads in Eq. 2-17 to the principle,

$$(X_c - X_1)^2 + (X_c - X_2)^2 + \ldots + (X_c - X_n)^2 \longrightarrow \text{minimum}$$

or the sum of the squares of the deviations from the "correct" reading must be a minimum. This *important result* is the basis of the widely used *least-squares method of plotting,* to be described in Chap. 8.

If we now differentiate Eq. 2-17 with respect to ΔP_{total} and X_c, we obtain

$$\frac{d(\Delta P_{total})}{d X_c} = -\left(\frac{\Delta x}{\sigma \sqrt{2\pi}}\right)^n \frac{1}{\sigma^2} [(X_c - X_1) + (X_c - X_2) + \ldots + (X_c - X_n)]$$

$$\times \exp\left[-\frac{(X_c - X_1)^2 + (X_c - X_2)^2 + \ldots + (X_c - X_n)^2}{2\sigma^2}\right] \quad (2\text{-}18)$$

There is only one way this derivative can equal zero. That is when

$$X_c = \frac{X_1 + X_2 + \ldots + X_n}{n} \quad (2\text{-}19)$$

and a test of the second derivative will show that this is a maximum point of the function. Thus we have proved that the most probable value of X_c is equal to the *numeric average of the n readings,* a proposition used without conscious thought by engineers and scientists the world over. We need not restrict the use of the numerical average to normally distributed density functions, but say only that for such a normal distribution the mean is the best estimate we have of the correct or true value.

Generally, if the error or deviation distribution is symmetric about a central value, the mean value of the readings is the best estimate of central tendency, even if the readings are not normally distributed. If, however, the parent population is *skewed* or *pathological*, as we shall consider later in this chapter, then the mean may not be the best measure of central tendency or the best reading.

Example 2-3 What is the best value of blowout current for the fuse distribution in Fig. 2-3, assuming that this distribution is a symmetric one?

Solution We form the mean from Table 2-1 or Fig. 2-3 by multiplying each reading by the number of readings in the segment, adding the totals for all the segments and dividing by the total number of fuses in the test (69).

$$[3(92) + 4(93) + 9(94) + 10(95) + 13(96) + 10(97) + 8(98)$$

$$+ 8(99) + 2(100) + 2(101)] \left(\frac{1}{69}\right) = 96.23 \text{ mA}$$

Thus our central-tendency value noted from Fig. 2-3 to be 96 mA is well verified numerically, suggesting that the distribution is reasonably symmetric. We shall study more specific tests for normality in Sec. 2-8.

2-7 PRECISION ERROR FROM SAMPLES

If we can now make the assumption that $X_c = \mu_x$, that is, that the sample mean is a good estimate of the population mean, then we can obtain the several x deviations by writing Eq. 2-14 as

$$x = X - X_c \tag{2-20}$$

When we use the sample mean, however, it can be shown by statistical theory that the *best estimate* of the sample standard deviation, symbolized by s, is given by

$$s = \left(\frac{\sum_{n=1}^{n} x^2}{n - 1}\right)^{1/2} \tag{2-21}$$

where the denominator $n - 1$, instead of n in Eq. 2-6, provides correction for the fact that the x values are computed from a *sample mean* and not the correct *population mean*. We shall define the *dimensionless deviation for a sample T* from

$$T = \frac{x}{s} \tag{2-22}$$

and assume that *T is a good estimate of* τ, based on σ in Eq. 2-9. Thus we can use Table 2-2 if we have a sample of data from which to obtain the standard deviation (SD). Equation 2-21 defines the SD found by all minicomputers having an automatic SD key.

When one wishes to find the sample standard deviation *s* without the intermediate step of obtaining the mean, the formula

$$s = \left[\frac{\sum\limits_{n=1}^{n} X^2 - \left(\sum\limits_{n=1}^{n} X \right)^2 \Big/ n}{n-1} \right]^{1/2} \tag{2-23}$$

is quick to use since it uses only the *actual readings* X_1, X_2, ..., X_n and not the deviations.

Table 2-2 shows that a deviation of 1 SD ($T = 1.0$) encloses about two-thirds of the distribution. Another useful measure of dispersion is the *probable error p*, defined as the deviation that encloses 50% of all the distribution. From Table 2-2 we see that this falls between τ values of 0.66 and 0.68. The actual relationship is

$$p = 0.675 s \tag{2-24}$$

Statisticians do not have much use for the probable error, noting that it does not have the general application of *s*. We shall show two uses for *p*. One is in making rapid estimates of statistical parameters as explained in Example 2-4, and the other is in dividing normally deviating data for rapid testing for the effects of treatments, as will be explained in Example 9-1.

Just as individual readings X_1, X_2, ..., X_n deviate around a mean with a dispersion measured by *s*, so we can imagine a distribution of *means* $X_{c,1}$, $X_{c,2}$, ..., $X_{c,n}$, where each mean is calculated from a sample of readings using Eq. 2-19. Even if all these samples are drawn from the same parent population, we expect that two different samples of *n* items each will not necessarily give the same X_c value. The question, then, is what is the *precision of the mean* of a set of *n* items; that is, how close to our computed sample mean X_c are we likely to find the population mean μ_x? If s_m is the standard deviation of the mean, it can be shown for symmetric distributions* that

$$s_m = \frac{s}{\sqrt{n}} \qquad (2\text{-}25)$$

or that the SD of the means of samples drawn from the same population is equal to the SD of the data divided by the square root of the number of data. Equation 2-25 shows that you will get twice as close to the true population mean with four readings as with one, and three times as close with nine readings.

Example 2-4 We are now ready to take a much harder look at the fuse data of Table 2-1 and Fig. 2-3. Let us find the standard deviation, the probable error, and then answer the two questions posed earlier: (a) What fraction of fuses will blow between 98 and 100 mA? (b) How many fuses should we expect to test before finding one that blows at 91 mA or less? Also, (c) What is the range of values to encompass the mean of the infinite population μ_x with only 1 chance in 20 of this true mean lying outside?

Solution We found the sample mean in Example 2-3 (96.23 mA), but Eq. 2-23 is so easy to use with data like those in Fig. 2-3 that we shall show its application.

$$\Sigma|X^2 = 3(92)^2 + 4(93)^2 + 9(94)^2 + 10(95)^2 + 13(96)^2 + 10(97)^2$$
$$+ 8(98)^2 + 8(99)^2 + 2(100)^2 + 2(101)^2 = 639{,}302$$

$$(\Sigma|X)^2 = [3(92) + 4(93) + 9(94) + 10(95) + 13(96) + 10(97) + 8(98)$$
$$+ 8(99) + 2(100) + 2(101)]^2 = 44{,}089{,}600$$

Then Eq. 2-23 gives

$$s^2 = \frac{639{,}302 - 44{,}089{,}600/69}{69 - 1} = 4.740$$

and

$$s = 2.18 \text{ mA}$$

*Young, 1962, pp. 92–96.

Equation 2-23 should be used only with a minicomputer, since a slide rule will not be sufficiently accurate.

The probable error p can now be found from Eq. 2-24.

$$p = 0.675(2.18) = 1.47 \text{ mA}$$

If 96.23 mA is the mean, then we expect 50% of the readings to be located between $96.23 + 1.47$, or 97.70 mA, and $96.23 - 1.47$, or 94.76 mA.

We can estimate the probable error by a simple *counting* procedure using Fig. 2-3. We note that the 96-mA segment contains 13 fuses, the 95-mA segment contains 10 fuses, and the 97-mA segment also contains 10, for a total of 33 fuses. Thus, roughly half the distribution is contained between the limits 94.5 mA and 97.5 mA. Then the histogram predicts a probable error of one-half of the $97.5 - 94.5$, or 1.5 mA, compared to 1.47 mA found more rigorously. This value would then give us a standard deviation (Eq. 2-24) of 1.5/0.675, or 2.22 mA, compared to the correct value of 2.18 mA. Thus one of the uses of the probable error is in making quick estimates of precision indicators by counting. It is always easiest to divide a histogram in half, as we did here, but this shortcut will give serious errors if only a few data are available.

(*a*) Continuing now, we wish to predict what fraction of fuses will fail between 98 mA and 100 mA. The value of the deviation x for 98 mA is $98.0 - 96.23$, or 1.77 mA. With a SD of 2.18 mA, T is 1.77/2.18, or 0.812 (Eq. 2-22). From Table 2-2 this is a P of about 0.584, or about 58.4%, of the fuses. We are interested in the part *above* the mean only, which contains 58.4/2, or 29.2%, of the fuses. The value of 100 mA is a deviation of $100 - 96.23$, or 3.77 mA, and its T value is 3.77/2.18, or 1.73. From Table 2-2 this is a P of about 0.916, or 45.8%, above the mean. Thus the fuses between 98 and 100 mA lie outside the probability area of 29.2% and inside the area of 45.8%, or involve a P of $45.8 - 29.2$, or 16.6%. We would thus expect, on the basis of this *normal law prediction,* to find 0.166 × 69, or 11.5, fuses within this range, or practically, 11 or 12 fuses. In fact, if we assign four of the eight fuses in the 98 segment in Fig. 2-3 to the range 98 to 98.5 mA, all eight to the 98.5-to-99.5 segment, and one of the two 100-mA fuses to the 99.5-to-100 mA range, then we have $4 + 8 + 1$, or 13, fuses, one more than the prediction of the normal model. Of course it is easy to get such good agreement with a sample of 69 items, a huge amount compared to most engineering tests where three or four replications of readings is usually the most one can expect. The point is that we can make identical predictions with four- or five-item samples, but with the expectation that such predictions are not likely to be so close.

(*b*) We wish to consider how many fuses we should test before we get a failure at 91 mA or less. This reading represents a deviation of −5.23 mA, and *T* (Eq. 2-22) is 5.23/2.18, or 2.42. Table 2-2 shows a probability of 0.984. The probability of finding a point outside the deviation, and in the lower tail only, is (1.0 − 0.984)/2, or 0.008. Since the chance of finding a reading below 91 mA is 0.008, which is the fraction 1/125, we expect that we would need to test, on the average, 125 fuses to obtain one in this lower tail, or almost twice as many as we did test.

(*c*) From Eq. 2-25 we see that the SD of the mean is

$$s_m = \frac{s}{\sqrt{69}} = \frac{2.18}{8.31} = 0.262 \text{ mA}$$

If we take two SDs on either side of the sample mean as the 95% probable range in which to find the population mean, we obtain (2 × 0.262) = 0.524. With the sample mean of 96.23 mA, we expect that the true population mean lies between 96.23 − 0.523, or 95.71, and 96.23 + 0.523, or 96.75, a total spread of about 1 mA. This is called a *95% confidence interval*, since we expect the chance of finding the true mean in this range to be 19 out of 20, or 95%.

2-8 THE PROBABILITY GRAPH

The fact that the theoretical distribution of Table 2-2 gave good predictions concerning the fuse distribution of Fig. 2-3 suggests that the two are similar in form. There are a variety of mathematical tests to determine the relative normality of a distribution, but we shall use the simplest: plotting the data on normal coordinates and observing the line that results. Normal coordinate paper can be purchased, but if you are caught short and want to make a rough plot, you can use regular graph paper with a horizontal line through the middle of the *y* (ordinate) axis marked 50%. Then beneath and equally spaced in descending order, label the *y*-axis intervals 38.8, 27.6, 19.8, 13.6, 7.9, 4.5, 2.4, and 1.2, all in percent. Above the 50% line, lay off eight more equally spaced intervals and in ascending order label these 61.2, 72.4, 80.2, 87.4, 92.5, 97.6, and 98.8, all in percent. Interpolating between these intervals is difficult, and proper paper gives a better presentation.

Let us prepare the fuse data of Table 2-2 for plotting on probability coordinates. Table 2-3 shows the extension of Table 2-1 to accomplish this, and Fig. 2-7 shows the first and last columns of this table plotted on probability

Table 2-3 Preparation of fuse (Fig. 2-3) data for normal coordinate plotting

Value (mA)	92	93	94	95	96	97	98	99	100	101
Number at value	3	4	9	10	13	10	8	8	2	2
Number at or below value	3	7	16	26	39	49	57	65	67	69
Fraction at or below value	0.043	0.101	0.232	0.377	0.565	0.71	0.826	0.942	0.971	1.0

coordinates. There are three tests of the sample distribution that can be made on this cumulative graph.

1. *Is the line straight?* The straighter the line, the more closely the data fit the normal assumption. Figure 2-7 suggests an excellent fit, as the various predictions in the previous section suggested.
2. *Does the 50% line pass through the distribution line at the mean value?* If not, the distribution is *skewed*, that is, has more items on one side of the mean than on the other. Figure 2-7 shows that whereas the computed mean is 96.23 mA, the value that divides the distribution in half on Fig. 2-7, what we might call the "graphical mean," is 95.75 mA. Thus the distribution is slightly skewed toward the low-current side of the mean. Put another way, 58% of the readings fall below the sample mean and 42% above.
3. *Does ±1 standard deviation occupy about 68% along the probability scale?* With a mean of 96.23 mA and a standard deviation of ±2.18 mA we have a 2s range of 94.05 mA to 98.41 mA. If we lay off this distance along the axis of Fig. 2-7, the distance along the probability axis is the difference between 88.5% and 22%, or 66.5%, compared to a value of 68% if the distribution were normally distributed. This is a good check on the slope of the line and again suggests a close fit.

Figure 2-8 suggests some of the ways in which random data can deviate from the normal model. *Skewness* was shown in a mild form on Fig. 2-7. When a distribution is very skewed, it is also likely to have curvature as in Fig. 2-8.

Flatness refers to a normal-type curve with a top that peaks too sharply or too bluntly. Such a curve might be symmetrical and pass test 2 above, but it would fail tests 1 and 3.

When the distribution curve or density function does not match these common forms, it is often called *pathological.*

Figure 2-9 shows a purely hypothetical example of the possible result of a hook gauge measurement in which the apparent water depth depends on whether the hook moves up or down to touch the water surface. Repeated, single-depth

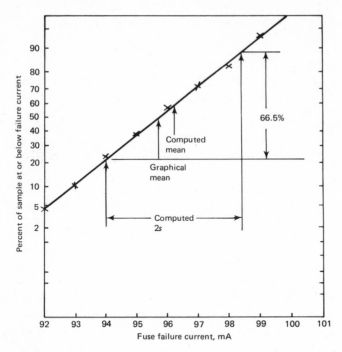

Figure 2-7 Fuse data from Table 2-3 plotted on normal coordinates to test the normality of the sample.

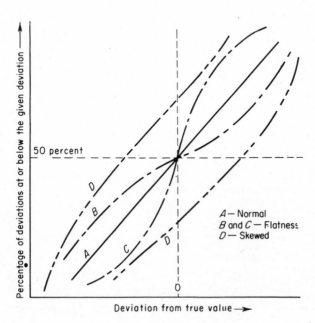

Figure 2-8 Sketch on probability coordinates showing possible plots for several types of nonnormal distributions.

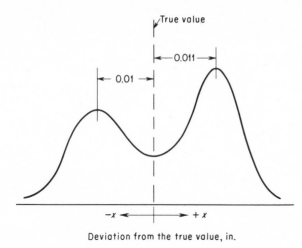

Deviation from the true value, in.

Figure 2-9 A hypothetical density function that might result from a large number of hook gauge readings as discussed in the text. The depth gauges of Example 2-2 would show similar curves if their readings were approached from both above and below the true reading.

calibration of the depth gauges in Example 2-2 might produce a similar shape, if the approach to depth were about equal from above and below. Clearly, such a density function will require many tests to define its shape. If the calibrator is not aware of the problem and happens to make the approach to the reading mostly from the same direction, he or she could never obtain the correct form of the distribution. Chapter 6 deals with experimental plans to avoid or detect such errors.

Example 2-5 An angle-factor integrator is an instrument used to find that fraction of the hemisphere over a solar panel that is blocked by some other object. Eleven students were provided with such a device and used it to obtain this proportion for a situation involving a wall and four solar panels, obtaining for the percent blockage by the wall, symbolized by $F_{1 \to 2}$, the following figures:[*] 13.6, 12.8, 13.5, 12.5, 13.0, 13.3, 13.5, 13.4, 13.8, 13.1, and 13.8. A computer analysis of the geometry gave a value of 13.87%. What can we say about this instrument with regard to accuracy and precision?

Solution The mean of this distribution is 13.3%. We can set up a table to find s using Eq. 2-21.

$F_{1 \to 2}$	0.136	0.128	0.135	0.125	0.130	0.133	0.135	0.134	0.138	0.131	0.138
$F_{1 \to 2} - 0.133$	+0.003	−0.005	+0.002	−0.008	−0.003	0.0	+0.002	+0.001	+0.005	−0.002	+0.005
$(F_{1 \to 2} - 0.133)^2 \times 10^6$	9	25	4	64	9	0	4	1	25	4	25

[*]For complete details on the experiment, see Schenck, 1970, pp. 79–88.

$$s^2 = \frac{\sum\limits_{n=1}^{n} x^2}{n-1} = \frac{1.7 \times 10^{-4}}{11-1} = 0.17 \times 10^{-4}$$

so that s is 0.0041, or 0.41%.

The mean of the 11 readings was 13.3% compared to the computer prediction of 13.87%; so the accuracy error in this set of determinations is about 0.6%, with the instrument reading too small a blockage figure. (The reasons for this bias error are given in the noted reference.) Two standard deviations equal almost 1%, so that we might define the worst precision error as ±1% at this condition. This figure, which is about 7% of the correct value of $F_{1\to2}$, is probably acceptable, since it is doubtful if the radiant emission characteristics of the solar panel or the blocking wall would be known to any greater precision.

Figure 2-10 shows the probability plot for the 11 data points. The data is more scattered since there are far fewer points than in the fuse case, but the line shown does not torture the data too much. Note that the main

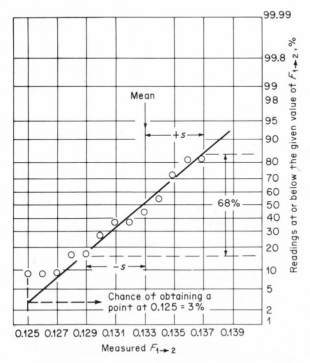

Figure 2-10 Probability plot of data in Example 2-6.

problem is with the reading at $F_{1 \to 2} = 0.125$. With the distribution shown by the line, this reading should occur only about 3% of the time, that is, about once in 33 measurements, whereas it appeared in this group of only 11 measurements. When drawing the "best" straight line on a scattered probability plot, it is the center points that should be given the most weight. In Fig. 2-10, the points at $F_{1 \to 2} = 0.125$ and 0.126 depend on a single datum.

We have just made the first test regarding the straightness of the line. Our second test, that the line pass through 50% at the mean, is quite well satisfied on Fig. 2-10. We can make the third test in another way than in Fig. 2-7 by showing what the data line would *predict* for the SD, based on plus-and-minus deviations about the mean that enclose 68%. From Fig. 2-10 we see that this graphically obtained value for s is just about 0.004, whereas the value found earlier from the sample was 0.0041, again a quite reasonable agreement for a distribution of 11 points.

2-9 ESTIMATING THE PRECISION ERROR OF AN INSTRUMENT

There is no way to look at an instrument and estimate its accuracy error. This can be done only by calibration. When we purchase an instrument, or take it off a crib shelf, we usually assume that it has no accuracy error, unless, of course, a calibration curve is attached.

In the case of precision error, which is *always* present at least to some degree, a rather rough-and-ready assumption is that "most" of the readings, say within ±2 SD, will fall within *one-half the least count of the instrument*. For example, the pressure gauge scale shown in Fig. 2-11 has a least count of 2.5 psi; so we would expect it to read consistently to within 1.25 psia of the input reading. Notice that the range of 0 to 10 psi is probably less precise than this, a common situation with inexpensive pressure gauges.

Figure 2-11 A typical inexpensive pressure gauge scale having a least count of $2\frac{1}{2}$ psi in the range 10 to 60 psi.

Sometimes a meter will be noted that reads to within "5% of full scale." If full scale is 100 V, then the error can be as much as 5 V, too much if one is checking a 9-V circuit.

Often, part of the precision error of an instrument reading stems from variations in the use of the instrument or associated equipment. A weighing tank may be timed as it fills with water, but some error in the timing may be introduced by operating the water valve or setting the weighing scale. A good way to check such measurements is to time two different quantities of water, say 100 s worth, then 200 s worth. If the large amount takes more (or less) time than twice the time taken for half the amount, there may be a bias error associated with starting or ending the measurement. The amount of the bias can be found from a series of runs by extrapolating to zero water, as we shall discuss in Chap. 7.

Sometimes the precision error in a measurement can be roughly estimated by watching a minute or so of *digital readout* of a fixed input reading. Whatever the reading, the digital output will often show continuous variation in the last one or two places, and it may be possible to estimate the range containing "most" of the output, then assume this to be about 2 SD.

Whatever the method—calibration with repeated readings, study of the scale, study of the output, or guesswork—we shall need a precision error figure for each test instrument if we are to combine these, as we do in Chap. 3, and obtain the expected precision of an overall experiment.

2-10 THE NORMAL DISTRIBUTION AS AN EXPERIMENTAL RESULT

We noted at the beginning of this chapter that the normal distribution is not only the most useful model of instrument error, but that many complete tests and experiments may produce sets of data that are normally distributed. When we can establish such a pattern, we often infer that we are dealing with a "natural" random variable, that is, a variable that is not responding to the parameters of the experiment but is simply being generated from the uncontrolled interior variations of the system. Of course, many modern experiments in natural energy generation and resource utilization often involve a search for a density function that describes the day-to-day values of wind, solar, or ocean current movement, or other naturally generated numbers. The normal distribution is by no means the only probability function used to describe these distributions, but it is always one that is "tried" on any distribution having a central tendency.

Figure 2-12 A blindfolded scuba diver attempts to point at a sound source 40 ft away. His buddy diver relates his pointing direction to compass direction and records the trial.

Figure 2-12 shows a reasonably complex experiment in which a disoriented scuba diver was asked to point toward an underwater sound source while his buddy diver hovered over him and noted the compass direction of his pointing, recording it to the nearest $10°$ (5, 15, 25, etc.) right or left. The purpose of the research is to seek the best sound format that will enable a buddy diver to locate a partner who is signaling an emergency.* Although a variety of sound patterns

*See Leggiere et al., 1970.

and head coverings were tried, nothing appeared to improve the very serious loss of sound directionality that humans suffer under water.

In all, 350 trials were made with six subjects; Fig. 2-13 shows a *normalized histogram* of the complete data. This figure contains about 95% of the points; the remaining 5% were scattered in the arc between ±110°, that is, the diver was pointing *away* from the sound source. To make a probability plot of this data, we assume that 2.5% of the backward-facing data belongs to the left-side data, and then accumulate percentages from the normalized histogram (Fig. 2-13), obtaining the results shown in Fig. 2-14.

Let us apply our three tests to this graph. First, we note that a straight line does *not* correlate the points. If you refer to Fig. 2-8, curve B, you will see that the distribution of Fig. 2-14 has a less acute form of *flatness*. Figure 2-14 shows that there are too many data points in the "tails" out beyond 75° both left and right.

Our data passes the skewness test in two ways. First we note that the average of the entire distribution is −0.8°, that is, almost zero. This is important, since any skewness of the distribution with respect to zero would suggest bias or cueing errors in the test setup of Fig. 2-12. Figure 2-14 shows that the mean of −0.8° divides the distribution almost in half, again suggesting that there is no skewness in the data. Here, then, is a case where the mean and the "expected" value are close together. Note, however, that when we compare the *mean* to zero, we are working with *both* the numbers of readings on each side of zero and

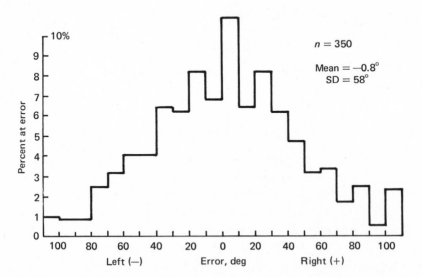

Figure 2-13 A normalized histogram using data from 350 pointing trials obtained as shown in Fig. 2-12.

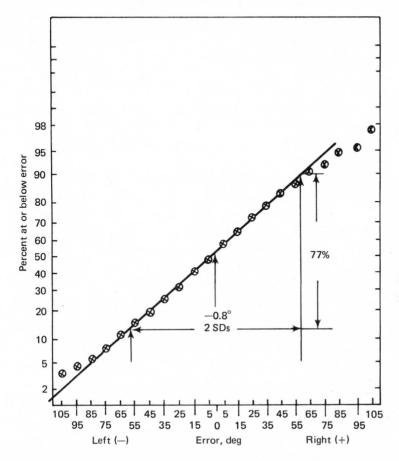

Figure 2-14 A plot on probability coordinates of the pointing data of Fig. 2-13. Note that there are too many points in the "tails."

their relative dispersion. When we compare the 50% point to −0.8° on Fig. 2-14, however, we only compare the *numbers* of data on either side and not how far away they are.

Figure 2-14 shows that the pointing data *fails* the third normality test. Two standard deviations enclose 77% of the sample, not the 68% we expect for normally distributed data. With this many data points such a large discrepancy is decisive proof that the distribution is *nonnormal.* Figure 2-14 shows that the standard deviation (58°) is too big and should be about 48° to match the line through the central points. This simply reiterates what we said in the straight-line test. There are too many points far from zero, producing the curvature on Fig. 2-14 and making the standard deviation too large.

There is no theoretical reason why scuba divers should have sound-beacon pointing errors that are normally distributed, but the existence of a nonskewed and yet nonnormal distribution suggested the following hypothesis: The divers could not explain how they determined the direction of the sound source but would often get hunches that they were doing well (or, sometimes, badly)—hunches that seemed to be reflected in the data. Divers pointing relatively closely to the correct direction may form a normal error sample, but when they are missing the target by 75 or 100°, they may actually have *no idea* where it is. In such a situation, they might be just as likely to point 10°, 130°, or even 180° away. Thus there may be *two different populations* of data making up Fig. 2-13: those in which the divers could hear and point more or less at the source, and those attempts, far fewer in number, where they simply pointed in a random direction. The combination of samples from two such groups would produce the effects we observe: tails too large and too far away from the mean. We shall see another example in Chap. 9 in which the failure to achieve a theoretical distribution leads to a hypothesis of two parent distributions.

The exact answer to the puzzle, however, is not the reason for this exercise. What we see is that the distribution model of the normal law can organize and make meaningful our calibration data, and even when it fails as it does in Fig. 2-14, it can yield clues to theories for the analyst and test engineer.

2-11 THE EFFECT OF NONNORMAL ERROR DISTRIBUTIONS

Figure 2-15 is a hypothetical error distribution of 50 length measurements so adjusted as to deviate following the normal law very closely. The sum of the squares of these 50 deviations is easily found:

$$\Sigma\, x^2 = 6(0)^2 + 10(0.2)^2 + 10(0.4)^2 + 8(0.6)^2 + 6(0.8)^2 + 4(1.0)^2 + 2(1.2)^2$$
$$+ 2(1.4)^2 + 2(2)^2 = 27.5$$

and from Eq. 2-21 the standard deviation is 0.749 in. From Eq. 2-24, $p = 0.675$ (0.749), so that the sample probable error in this case is 0.506 in. Checking Fig. 2-15a, we note that an x deviation of ±0.5 in encloses a little more than half the total sample, or 26 items.

Now consider Fig. 2-15b, which is an example of the so-called roulette wheel distribution, so named because every deviation is equally probable, just as every number on an honest roulette wheel is equally probable. This distribution is

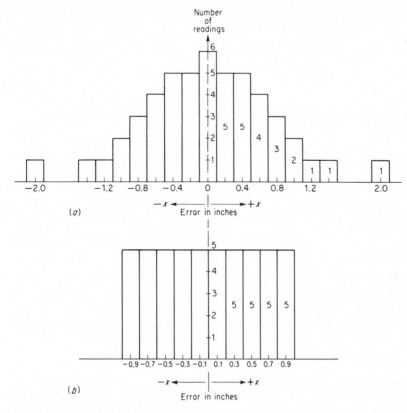

Figure 2-15 Histograms of 50 errors distributed (*a*) as normal and (*b*) as a "roulette wheel" distribution.

adjusted to contain 50 readings and to have a probable error of ±0.5 in as before. Computing the sum of the squared deviations again,

$$\Sigma x^2 = 10(0.1)^2 + 10(0.3)^2 + 10(0.5)^2 + 10(0.7)^2 + 10(0.9)^2 = 16.5$$

Equation 2-21 now gives a standard deviation of 0.58 in compared with the value of .749 in found for the case in Fig. 2-15*a*.

Suppose we were to assume that the distribution of Fig. 2-15*b* were normal, perhaps because we had only a few readings instead of the 50 shown. Following this assumption, we might obtain the probable error from Eq. 2-24, which is based on the normal law equations. This equation would predict a probable error of 0.392 in instead of the correct value of 0.5 in, which is easily obtained by inspection of Fig. 2-15*b*.

It should be obvious that few real error distributions will deviate from normality as drastically as Fig. 2-15*b*. Yet even in this extreme case, the (incorrect) assumption of normality leads to a prediction of the plus-or-minus deviation enclosing half the readings, which is only a little over 20% lower than the correct value. A single example is not sufficient to prove a principle. Nevertheless, the implication of this brief analysis—that large departures from normality do not lead to important mistakes—is amply borne out in practical experimentation. The problem is not, as some statisticians and other specialists would have one believe, that a model or method may be used clumsily or in the wrong place. Far more serious in the present state of engineering experimentation is that the statistical model will not be used at all.

One precaution might be appended to these rather optimistic conclusions concerning the application of the normal model. When an engineer is interested in extreme-value situations, crude or superficial model selections can lead to serious mistakes or real disasters. A civil engineer who estimates that a given stream reaches a particular extreme and unlikely flood level only once in 100 years may be embarrassed when the level turns out to occur on an average of once every 5 years. Such questions deal with the tails of statistical distributions and great care must be exercised when extreme values may produce dangerous and destructive occurrences.

PROBLEMS

2-1 Replot the data from Example 2-1 in a deviation plot as in Example 2-2. Would you say that the CO_2 meter has a sum-type or a product-type error?

2-2 Make a parity graph of the data shown in Example 2-2 and discuss the two calibration curves that result.

2-3 In Example 2-1, we noted that a third way of plotting calibration data was to plot the percent error, defined as the deviation error divided by the correct reading versus the correct reading. Make such plots for the calibration data of Examples 2-1 and 2-2. What form does the calibration data of an instrument giving a sum-type error take in such a plot? What form is produced when the instrument has a product-type error?

2-4 We noted in discussing the deviation plot (Fig. 2-2) that this could not be easily used to correct gauge readings to true readings. Replot the deviation data of Example 2-2 for each gauge against the gauge depth rather than the true depth. Compare this calibration plot with the parity graph type of calibration curve (Fig. 2-1).

2-5 The accompanying graph shows the data for a single calibration test on a pressure gauge giving five test points. Make a deviation plot of this data and comment on the kind of error suffered by the gauge.

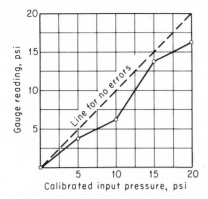

2-6 The accompanying figure shows the results of four separate calibration tests on the pressure gauge of Prob. 2-5. Draw the best calibration curve through these points and comment again on the kind of accuracy error suffered by the gauge.

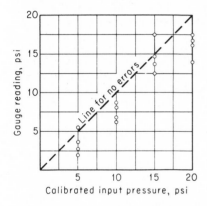

2-7 Find the standard deviation at each of the four calibration points for the graph in Prob. 2-6 and then plot this, representing the variability at each pressure, against the true pressure. What does the graph suggest about the precision-error behavior as a function of pressure?

2-8 A furnace temperature probe is found to have a normally distributed precision error with an η value of $0.04°C^{-1}$ at $1000°C$. If a sample of 20 readings is made at this setting, how many do we expect to fall between the limits of $990°C$ and $1010°C$? What is the value of the standard deviation for this instrument, and what is the range enclosing (2 s) deviations, that is, about 95% of the data?

2-9 Several student groups use a Brinell hardness testing machine on specimens of known and unknown hardness. The following deviations from the correct indenter hole diameter occur: 1 of deviation -0.20 mm, 1 of -0.10 mm, 4 of -0.05 mm, 13 of zero deviation, 7 of deviation $+0.05$ mm, 4 of deviation $+0.10$ mm, and 1 of deviation $+0.20$ mm. Find the index of precision of this sample and the standard deviation, and make a probability plot of the data; then apply the three normality tests.

2-10 A second run using 35 of the same kind of fuses as in the data of Table 2-1 was made with a faster rate of current increase than in the previous experiment. The results were as follows:

Blow current (mA)	93	94	95	96	97	98	99	100	101	102	108
Number of fuses	1	1	6	4	7	7	5	0	1	2	1

Find the important statistical parameters for this distribution and comment on its normality. Compare the distribution with that of the 69-fuse sample in the text. Notice that one fuse in this sample did not blow until 108 mA. Assuming that this distribution is normal, how many fuses would we expect to test before having one last to 108 mA?

2-11 A hook measures water depth, and its readings are found to be normally distributed with an η value of 3.1 m^{-1}. If a large number of determinations were made, one-quarter of them would fall outside what range of depths if the "true" value is 0.46 m?

2-12 Twelve readings are made with a sextant as follows: all 22°, plus 30', 40', 40', 10', 30', 20', 0', 30', 60', 20', 40', 30'. Decide if this is drawn from a normal distribution and find the mean standard deviation. If this were a noon sight, and latitude error in nautical miles is equal to the error in minutes in the sight, what is the maximum (20-to-1 odds) position error in latitude expected from the instrument?

2-13 A Geiger counter error distribution has 25% of its readings one count too low, that is, at an x value of -1.0 count. Fifty percent of its readings are at x of zero (correct number of counts), and 25% of the readings are at an x value of $+1.0$ (one count too high). Sketch the probability plot for this distribution and find the standard deviation. What is the chance of getting three consecutive readings that are either exactly correct or smaller than the correct number of counts?

2-14 Sketch the form of curve you would expect to obtain on probability coordinates for the following nonnormal distributions:

(*a*) A normal distribution with the lower tail clipped off, as might occur with an instrument with a zero stop
(*b*) A distribution that gives a histogram in the shape of an equilateral triangle

2-15 An ohmmeter measures a standard 10,000-Ω resistor a number of times. One-half of all the readings lie outside the range 9,850–10,150 Ω. Estimate the modulus of precision for the instrument assuming a normal distribution. Find the percentage error on this range, defined here as the probable error divided by the nominal value (10,000 Ω).

2-16 Make a probability plot of the data shown in Fig. 2-15*b* and discuss its application to the three rules for normality given in the text. Compare this presentation to the acoustic scuba plot, Fig. 2-14, and discuss the differences in terms of deviations from the normal ideal.

2-17 An optical pyrometer is sighted on a fixed temperature source, and the results are as follows:

Temperature (K)	1850	1900	1950	2000	2050	2100
Number of readings	1	9	6	18	10	2

Decide if the sample seems normal and find the important statistical parameters of the distribution.

2-18 The average annual birth date of an insect pest is noon on day X. Half the specimens studied fall within ±36 h of this time, and the distribution is about normal. A pesticide has a lethal duration of 2 days and a kill probability of 98% during this period. How few applications are necessary to be certain of killing 95% of the pests? At what time, in relation to noon of day X, should the first application occur?

2-19 The variation of a 2.5-cm shaft is found to be normally distributed, with half the samples falling within ±0.0125 cm. The variation of bearing-hole diameter (around the nominal 2.5 cm) is also normal, with half the values falling within 0.00625 cm. Assuming that we hand-fit by trial as many shafts and bearings holes as we can, how many of the 50 shafts will not fit?

2-20 Given the situation in Prob. 2-19, what nominal bearing-hole diameter (instead of 2.5 cm) should we specify such that 100 parts can be mated by trials?

2-21 The *mean deviation,* or *average deviation,* α is a statistic measured by the sum of the absolute values of the individual deviations divided by the number of deviations or

$$\alpha = \frac{\sum_{n=1}^{\infty} |X_c - X_n|}{n}$$

It can be shown for a normal distribution that the sample SD, s, is related to the sample mean deviation a by $s = 1.25a$. What is the probability of finding a reading within $\pm a$? Using the data for the fuse test (Table 2-1), find the actual sample value of the mean deviation and compare its predicted probability with the actual number of points enclosed by $\pm a$.

2-22 In Example 2-5 we suggested that the mean of the 11 angle-factor readings, 13.3%, was actually biased low compared to the true value of 13.87%. The SD of the group of readings is 0.41%. Can we say that the noted bias is statistically established using the SD of the mean?

2-33 The scuba pointing data in Fig. 2-12 had 350 data points and a SD of 58°. By what amount might the expected mean of zero vary from zero and still be within a 68% chance of expectation, assuming that we can treat this as a normal distribution?

2-24 Two means are calculated from their respective samples and they have the same SD of the mean s_m. If the two means are exactly $2s_m$ apart, what is the chance that *both* population means lie between the two sample means?

2-25 A series of tests on automatic parachute-opening devices for dropping equipment gives a SD of 67 m. What fraction will smash their load if deployment *must* occur at least 25 m from the ground and the altitude setting is 150 m?

2-26 The following data resulted from a series of student determinations of Rockwell hardness. For specimen A, 97.0, 98.7, 99.9, 99.5, 97.1, 99.5, 92.0, 100.6, 99.7, 98.0, 98.5, 99.5, 99.7, 99.5, 99.0, 98.5, 99.5, 98.8, 98.5, 99.1, 98.4, 96.6, 97.2, 101.7, 97.2, 98.2, 97.5, 97.7, 99.0, 99.0, 97.5. For specimen B, same machine and groups, 85.6, 87.1, 87.9, 86.9, 85.6, 85.2, 85.5, 85.7, 84.7, 86.4, 80.0, 85.0, 82.0, 86.0, 86.0, 87.3, 84.5, 87.0, 87.3, 85.4, 91.0, 90.0, 90.8, 89.2, 91.0, 90.4, 84.1, 81.7, 87.4, 87.4, 84.0, 85.2. Find the average of each distribution, and use this to plot on probability paper the distribution of deviations from the average. From this plot, *estimate* the probable error and the standard deviation for each distribution. (*Note*: The data may be rounded to the nearest 0.5 hardness number.)

2-27 Compute the standard deviation and the probable error of the machine for the two specimens in Prob. 2-26, given that the higher the Rockwell number, the harder the specimen. Would you say that this hardness tester was more or less accurate on harder surfaces?

2-28 Assuming that the average of the Rockwell data on specimen B of Prob. 2-26 is correct, select in a random manner 1, 4, 9, 16, and 25 readings from the list. Show whether the theorem relating to the precision of an average increases as the square root of the number of readings is revealed in this case. (*Hint*: Plot on log paper the difference between the mean reading and the average of sets of 1, 4, 9, 16, and 25 items versus n. Ideally, what shape should this curve have?)

2-29 A precision sports car odometer is checked over a measured mile and found to have a SD of 1% around the true value. In a 100-mile run, estimate the uncertainty in the mean and give the maximum and minimum distances traveled that include 95% of all such runs.

2-30 It is estimated that the probable error in daily kilowatt (kW) output from a wind machine is .7 kW with a mean of 2.2 kW over a large number of days. Assuming that the distribution is reasonably normal and that it requires at least 150 W to operate the auxiliaries so that the machine has zero output below 150 W, what is the probability of finding two consecutive days in which there was zero power? How many times a year will this occur?

2-31 Suppose that the right side of Fig. 2-15a is reversed with the left side and the extreme values moved in to give the following double-peaked and symmetric error distribution:

x (in)	−0.14	−1.2	−1.0	−0.8	−0.6	−0.4	−0.2	0	+0.2	+0.4	+0.6	+0.8	+1.0	+1.2	+1.4
No.	3	5	5	4	3	2	2	2	2	2	3	4	5	5	3

Sketch a histogram of this distribution and estimate by inspection the approximate $\pm x$ value that encloses one-half the reading errors. Now compute, using all 50 items, the standard deviation of the sample. Suppose that one were to assume this a normal distribution and use this standard deviation to predict the $\pm x$ range that will enclose 50% of all the errors. What would be the value of this predicted x range, and how would it compare with that found from the density function itself?

2-32 A weighted box is used to find the coefficient of friction between a piece of tire and an asphalt road, obtaining from five trials 0.88, 0.88, 0.86, 0.84, and 0.87. An automobile at the same location was used in braking tests to obtain the same coefficient, obtaining in four trials 0.87, 0.88, 0.87, and 0.87. Discuss the question of whether the means of these two samples are probably the same.

2-33 The mean length of sun per day in July in a given area is 3.2 h with a SD of 1.6 h. With a given solar cell layout, we can charge a battery in July in 4 h. In how many of the 31 days of July can we hope to recharge the battery completely in one day? What is the "worst," that is, the day of least charge, that we might expect out of the 31? What is the chance that the day following this worst day will be sufficient to complete recharging the battery, assuming that recharging started on the worst day?

2-34 A concrete-block machine makes the following blocks, all masses in kilograms: 25.0, 25.2, 25.2, 25.0, 25.5, 24.8, 24.6, 24.9, 24.8, and 25.0. What is the probable range of the population mean block mass made by the machine that will include about 95% of the possible estimates? How many more blocks would we have to test with this same SD result to establish that the population mean lies between 24.95 and 25.05 kg, assuming that the mean remains at 25.0 kg?

2-35 The test on underwater sound location noted in Sec. 2-9 actually consisted of several groups of tests involving different sound frequencies in the water. The data made at 600 hertz (hz) gave the following distribution of 28 pointing errors (negative is left, positive is right, facing the target), all in degrees: −55, −45, −35, −25, −15 (two), −5, +5, (six), +15, (three), +25, +35 (four), +45, (three), +55 (two), +75, +85. Show a histogram and normal plot, and find the SD, mean, and standard error of the mean. Compare this sample distribution with Fig. 2-13. Sketch the Fig. 2-14 line on your probability plot and discuss (a) the normality of this sample, (b) the question of whether this sample mean can be considered to be drawn from the same population as the 350-sample mean (−0.8°) or else zero, and (c) whether the spread of the two distributions is similar.

2-36 The mean failure depth for deep-ocean bouyancy floats is 7800 m with a SD of 470 m. We obtain a failure at 7000 m. If the distribution is normal, estimate how many floats we used before this failure.

2-37 In an experiment on maximum exhale pressure drop in a tube, the following data result. Each L/D point was replicated 10 times. How does the error of this experiment behave as a function of increasing length? Plot a graph of L/D versus Δh and show the 95% confidence lines on the plot.

L (ft)	L/D	h (in H_2O)	
5.5	88.0	0.05	0.06
		0.05	0.04
		0.05	0.05
		0.06	0.05
		0.06	0.05
		Avg. = 0.052	
5.0	80.0	0.05	0.05
		0.06	0.04
		0.06	0.06
		0.05	0.04
		0.05	0.05
		Avg. = 0.051	
4.0	64.5	0.06	0.04
		0.05	0.05
		0.05	0.05
		0.05	0.04
		0.05	0.04
		Avg. = 0.048	
3.0	48.0	0.05	0.04
		0.05	0.05
		0.04	0.05
		0.04	0.05
		0.04	0.05
		Avg. = 0.046	
2.0	32.3	0.04	0.048
		0.044	0.044
		0.044	0.048
		0.042	0.043
		0.048	0.048
		Avg. = 0.045	
1.0	16.0	0.02	0.016
		0.018	0.016
		0.02	0.02
		0.02	0.022
		0.02	0.02
		Avg. = 0.0188	
0.5	8.0	0.016	0.016
		0.008	0.01
		0.012	0.006
		0.008	0.006
		0.01	0.01
		Avg. = 0.0096	

BIBLIOGRAPHY

Baird, D. C.: *Experimentation: An Introduction to Measurement Theory and Experimental Design,* chaps. 2 and 3, Prentice-Hall, Englewood Cliffs, N.J., 1962.

Beers, Y.: *An Introduction to the Theory of Error,* Addison-Wesley, Reading, Mass., 1957.

Boonshaft, J. C.: Measurement Errors: Classification and Interpretation, *Trans. ASME,* vol. 77, no. 4, pp. 409–411, May, 1955.

Burington, R. S.: Handbook of Mathematical Tables and Formulas, McGraw-Hill, New York, 1949.

Cramer, H.: *The Elements of Probability Theory,* part I, Wiley, New York, 1955.

Davis, H. E., G. E. Troxell, and C. T. Wiskocil: *The Testing and Inspection of Engineering Materials,* 3rd ed., McGraw-Hill, New York, 1964.

Fisher, R. A.: *Statistical Methods for Research Workers,* chap. 3, Hafner, New York, 1954.

Hald, A.: *Statistical Theory with Engineering Applications,* chaps. 5, 6, and 7, Wiley, New York, 1952.

Leggiere, T., J. Van Ryzin, J. McAniff, and H. Schenck: Sound Localization and Homing of Scuba Divers, *Mar. Tech. Soc. J.,* vol. 4, no. 2, pp. 27–34, 1970.

Palmer, A. deF.: *The Theory of Measurements,* McGraw-Hill, New York, 1930.

Parratt, L. G.: *Probability and Experimental Errors in Science,* Wiley, New York, 1962.

Schenck, H.: *Case Studies in Experimental Engineering,* McGraw-Hill, New York, 1970.

Spiegel, M. R.: *Probability and Statistics,* Schaum's Outline Series, McGraw-Hill, New York, 1975.

Tippett, L. H.: *Technological Applications of Statistics,* chaps. 8 and 9, Wiley, New York, 1950.

Wilson, E. B.: *An Introduction to Scientific Research,* chap. 9, McGraw-Hill, New York, 1952.

Worthing, A. G., and J. Geffner: *Treatment of Experimental Data,* chaps. 6 and 7, Wiley, New York, 1943.

Youden, W. J.: Systematic Errors in Physical Constants, *Phys. Today,* vol. 14, no. 9, pp. 32–43, September, 1961.

Young, H. D.: *Statistical Treatment of Experimental Data,* McGraw-Hill, New York, 1962.

THREE

ERROR AND UNCERTAINTY
IN COMPLETE EXPERIMENTS

In the previous chapter we studied the kinds of error that can occur in the measurement of a single quantity by an instrument system comprised of a sensing element, a secondary or indicating portion, and an observer. The majority of engineers are concerned with experimental systems in which several instruments are reading several quantities, and these measurements must be combined through some mathematical process to yield a final result. That such a situation may be fraught with difficulty should be evident. Measurements that in their raw form appear quite accurate may turn out at the end of a computational chain to possess errors that virtually destroy the validity of the test. Or, if luck is with the investigator, seriously imprecise data may play so small a role in the final analysis that their questionable nature can be ignored. But there is no need to count on luck, for it is usually possible to investigate the matter of result accuracy thoroughly before a single piece of test apparatus is erected or even purchased.

As soon as we deal with the readings, errors, or uncertainties of more than a single instrument, we introduce the problem of considering the various instruments on a common basis, as far as their error or uncertainties are concerned. In this chapter we shall assume that all instruments involved suffer from either *precision error* or uncertainty, which we treat as precision error. In Chap. 2 we saw that we could express this error or uncertainty as having a *distribution* (probably normal in form) and as being described by a *precision index* (such as the standard deviation, probable error, or modulus of precision). When we are to combine the errors or uncertainties of two or more instruments, we must, of course, put them all on the same basis.

3-1 THE PRECISION INDEX
OF A PRODUCT OR QUOTIENT

As we shall see in the following chapter, one of the most common types of functions found in experimental work is combinations of *products* and *quotients* (the dimensionless numbers). Typical examples are the familiar Reynolds number (velocity times a length times a density divided by viscosity) or the Mach number (vehicle velocity over speed of sound). Consider a general result R that is a function of the product of two measured quantities X and Y,

$$R = kX \cdot Y \tag{3-1}$$

where k is some constant that we assume to be known exactly. Now assume that a sample of X readings shows a standard deviation s_x and that the Y readings show an s_y deviation. If x_1 is the given deviation from X_c because of the precision error of the x measurement, and y_1 is the deviation of the Y measurement occurring at the same time, Eq. 3-1 for this specific pair of readings (out of a sample of n such pairs) becomes

$$R_c + r_1 = k(X_c + x_1)(Y_c + y_1) \tag{3-2}$$

where r_1 is the deviation of the result and the subscript c refers, as in Chap. 2, to the correct or average reading. Equation 3-2 becomes

$$R_c + r_1 = k(X_c Y_c + x_1 Y_c + X_c y_1 + x_1 y_1) \tag{3-3}$$

where the term $x_1 y_1$ is of second order and may be ignored. Then, Eqs. 3-1 and 3-3 yield

$$r_1 = k(x_1 Y_c + y_1 X_c) \tag{3-4}$$

Similarly, for other pairs of X and Y values,

$$r_2 = k(x_2 Y_c + y_2 X_c) \tag{3-5}$$

and so on. From the definition of s in Eq. 2-21, we see that

$$s_r^2 = \frac{\Sigma \, r^2}{n - 1} \tag{3-6}$$

and from the n equations, 3-4, 3-5, and so forth,

$$\Sigma r^2 = [Y_c^2 \ \Sigma x^2 + X_c Y_c \ \Sigma(xy) + X_c^2 \ \Sigma y^2] \tag{3-7}$$

We assume that the term $\Sigma(xy)$ equals zero because any particular product of x and y is as likely to be positive as negative and the summation of a large sample of such products will tend to zero. Then substituting Eq. 3-7 in Eq. 3-6 yields

$$s_r^2 = k^2 \left(Y_c^2 \ \frac{\Sigma x^2}{n-1} + X_c^2 \ \frac{\Sigma y^2}{n-1} \right) \tag{3-8}$$

Again applying the definition of s from the previous chapter and inserting Eq. 3-1, we obtain finally

$$\frac{s_r^2}{R_c^2} = \frac{s_x^2}{X_c^2} + \frac{s_y^2}{Y_c^2} \tag{3-9}$$

which is directly applicable, as will be shown in an example. Following the same method as outlined here, it is easily proved that Eq. 3-9 will hold for the case

$$R = \frac{X}{Y} k$$

so that when

$$R = \frac{XY}{Z} k$$

the expression

$$\frac{s_r^2}{R_c^2} = \frac{s_x^2}{X_c^2} + \frac{s_y^2}{Y_c^2} + \frac{s_x^2}{Z_c^2} \tag{3-10}$$

will apply. Notice that the term s_c/R_c represents the percentage of the correct reading represented by the standard deviation and is thus a type of *percent error.* Thus Eq. 3-10 is a general mathematical form of the rule: When the result is a function of quotients and/or products of a series of measurements, the square of the percent error of the result is equal to the sum of the squares of the percent errors of the individual measurements.

The foregoing analysis is not restricted to the use of the standard deviation or to instruments that deviate only in a normal manner. Kline and McClintock have extended the method to include a general deviation limit $\pm w$, where w might enclose 68, 50, 95%, and so on, of all the readings of a given instrument.[*] If $\pm w$ encloses 95% of all readings, Kline speaks of 20-to-1 odds that a reading will fall outside this limit. For this general deviation, Eq. 3-10 becomes

$$\frac{w_r^2}{R_c^2} = \frac{w_x^2}{X_c^2} + \frac{w_y^2}{Y_c^2} + \frac{w_s^2}{Z_c^2} \tag{3-11}$$

We must, of course, be sure that each w encloses the same percentage of its total population and, furthermore, that each of the several instruments deviates in a symmetrical manner. If a distribution is skewed, the terms Σxy in Eq. 3-7 will not, in the limit of $n = \infty$, be zero. Another restriction on Eqs. 3-10, 3-11, and most of the equations yet to come in this chapter is that precision errors or uncertainties, w_x, w_y, w_z, be independent.[†] For example, the index of refraction N of a prism as a function of two measured angles A and B may be given as

$$N = \frac{\sin \, [(A + B)/2]}{\sin \, (A/2)}$$

It is a temptation to treat the problem of finding the uncertainty or error in N as a function of the quotient of two quantities, $\sin \, [(A + B)/2]$ and $\sin \, (A/2)$, but this would be incorrect because these two quantities are not independent. Thus Eqs. 3-10 or 3-11 cannot be used, and we must obtain a more general expression for error propagation capable of handling sines, logs, and so on.

Example 3-1   Figure 3-1 shows a device intended to obtain the coefficient of friction μ for rubber auto tire samples on various surfaces. A constant-force piston presses the tire sample against the "road" (Fig. 3-2), so the energy loss ΔE, as the tire sample rubs past the surface over the interference distance d, is found from $\Delta E = \mu N d$, where N is a constant force of 27.24 N and is equal to the interior head pressure P, times the piston area A. This loss of energy at the surface must be reflected in the loss of height of the piston of mass 2.5 kg from h_4, the height with no contact, minus h_3, the height when the tire is in contact, or

$$\Delta E = Mg(h_4 - h_3)$$

[*]Kline and F. McClintock, 1953.
[†]Worthing and Geffner, 1943, pp. 210–212.

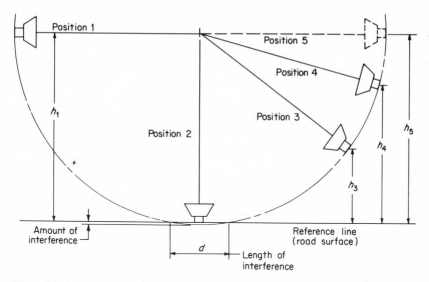

Figure 3-1 Swinging pendulum apparatus for measuring the coefficient of friction between a rubber sample and a "road" surface.

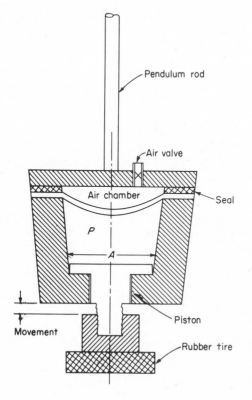

Figure 3-2 Piston head of Fig. 3-1 apparatus for maintaining constant pressure on road during tangent contact.

Equating the energies gives

$$\mu = \frac{Mg(h_4 - h_3)}{Nd}$$

The students examined the constant-force head and concluded that its inside pressure and mass did not change and thus that N and M had no precision error. They estimated that they could read $h_4 - h_3$ to within about 2.5 mm (0.1 in), which we shall take to be 2 SDs. The interference distance d was taken as a nominal 68.5 mm with an uncertainty from one run to the next assumed to be about 10%, which again we shall assume is about a $2s$ deviation. Predict the error in M and compare with the following set of five readings made under these conditions:

$h_4 - h_3$ (mm)	17.65	21.18	19.81	23.29	24.71
μ	0.25	0.30	0.281	0.33	0.35

Solution Let us assume a nominal value for $h_4 - h_3$ of 20 mm for our error estimate. Then, from Eq. 3-11 with w_μ the uncertainty in μ,

$$\frac{w_\mu^2}{\mu^2} = 0.1^2 + \left(\frac{2.5}{20}\right)^2 = 0.0256$$

and the uncertainty ratio in μ is predicted to be 0.16, or 16%. If the five experimental points are used with Eqs. 2-19 and 2-21, we obtain a mean of 0.303 for μ and a sample standard deviation of 0.0395. If our estimates were based on "most" of the data, say $2s$ deviations, then our data show a percent deviation of 2(0.0395/0.303), or 26%.

Evidently we did not allow enough uncertainty in either the interference (d) or the measurement $h_4 - h_3$. We might compute the uncertainty ratio in d again from Eq. 3-11, but now taking the experimental value for $2s$ for μ,

$$\left(\frac{w_d}{d}\right)^2 = 0.26^2 - 0.0156 = 0.052$$

and $w_d/d = 0.228$. This suggests that the uncertainty (or variation, it does not matter which) in d must be about ±22.8% of the nominal value of 63.5 mm, this including most of the readings to account for the experimentally found standard deviation. But is it reasonable that this contact should vary by almost a quarter of its value? Evidently we have two error-generating measurements, and the data give us no way to look at their effects independently.

Note, however, that if we were to run the apparatus at a different value of μ, the absolute magnitude of $h_4 - h_3$ would change, but the contact distance d would not. The students tried out the apparatus on a rough asphalt road (the first test was on smooth, painted, lab-floor concrete). Their eight determinations gave for μ: 0.89, 0.84, 0.85, 0.80, 0.83, 0.79, 0.90, 0.85, with a mean of 0.844 and a sample SD of 0.0385.

Now, looking back, we see that the SD of the low-μ test was almost the same, 0.0395. The percentage error in μ based on $2s$ in the low-μ test was 26% and in this high-μ test is only 2 (0.0385/0.844), or 9.1%.

Equation 3-11 shows that any random error in the contract distance d must contribute an equal percentage error at both the low- and high-μ tests, since its nominal value is unchanged and there is nothing in the high-μ test that would change its random error. Thus the d measurement could not have contributed 22.8% error at the low-μ test, since it would have had to do the same at the high-μ test, which had only 9.1% error in total. If, however, $h_4 - h_3$ had an uncertainty of 5.32 mm (0.21 in) in the low-μ test, an uncertainty of 5.22 mm (0.204 in), in the high-μ test, this would completely account for all the random variation in μ.

The contact distance d may have accuracy error or bias, but we can detect no random error whatever in its measurement. Note that the apparatus is far more precise in high-μ tests than at low. If we were to run the apparatus on pavements of varying μs, we would obtain an *error band* or *scatter band* (Fig. 3-3) that would enclose most of the data. What statisticians call a *confidence interval* can be shown by connecting the points $2s$ above the line and $2s$ below the line. In Fig. 3-3 we show this confidence interval as a dotted line because we do not know if the interval is a constant width for all μ values. We expect that about 95% of all points taken with the machine will fall within these confidence lines.

Completing this analysis, we should note that we *have not calibrated* the apparatus, since we have no idea of the "true" friction coefficients for the floor or asphalt. We have, however, gone far toward establishing its random error behavior. And most important, we have shown that to improve the instrument, we must work first with the scales and pointer that read h_4 and h_3.

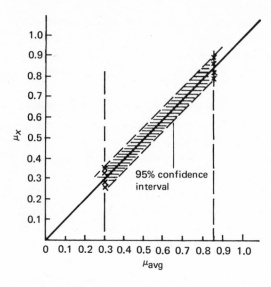

Figure 3-3 Confidence interval graph showing envelope of points obtained from apparatus of Fig. 3-1. Since this was not a calibration, we do not know where this interval lies in relation to the "true" values of the friction coefficient.

3-2 FINDING PRECISION INDEXES OF A GENERAL FUNCTION

We could continue, as in Sec. 3-1, finding special rules for a variety of special cases, that is, sums and differences, powers, logarithmic equations, trigonometric functions, and so on. But this would by no means complete our list, since we might find these classes of functions combined so that even more specialized and complex rules would be needed. Instead, let us develop a general method whereby any function can be analyzed to give the error in the result. Consider the general case of a result R, which is a function of the two measured variables X and Y:

$$R_c + r_1 = f(X_c + x_1, Y_c + y_1) \tag{3-12}$$

If this function is continuous and has derivatives (and it is difficult to imagine an experimentally determined function that fails such criteria), we can expand it in a *Taylor series*, using the first two terms only:

$$R_c + r_1 = f(X_c, Y_c) + \left[\left(\frac{\partial R}{\partial X_c} \right)_y \frac{X_c + x_1 - X_c}{1!} + \left(\frac{\partial R}{\partial Y_c} \right)_x \frac{Y_c + y_1 - Y_c}{1!} \right]$$

Or, since $R_c = f(X_c, Y_c)$,

$$r_1 = \left(\frac{\partial R}{\partial X_c} \right)_y x_1 + \left(\frac{\partial R}{\partial Y_c} \right)_x y_1 \tag{3-13}$$

where, as before, the lowercase letters apply to deviations from the correct readings, and

$$\sum r^2 = \left(\frac{\partial R}{\partial X_c}\right)_y^2 \sum x^2 + 2\left(\frac{\partial R}{\partial X_c}\right)_y \left(\frac{\partial R}{\partial Y_c}\right)_x \sum xy + \left(\frac{\partial R}{\partial Y_c}\right)_x^2 \sum y^2$$

Again, $\Sigma\, xy$ tends to zero and $s_c^2 = \Sigma\, r^2/n$, so that

$$s_c^2 = \left(\frac{\partial R}{\partial X_c}\right)_y^2 s_x^2 + \left(\frac{\partial R}{\partial Y_c}\right)_x^2 s_y^2 \tag{3-14}$$

and for the uncertainty interval w,

$$s_r^2 = \left(\frac{\partial R}{\partial X_c}\right)_y^2 w_x^2 + \left(\frac{\partial R}{\partial Y_c}\right)_x^2 w_y^2 \tag{3-15}$$

and so on for any other precision indexes.

3-3 APPLICATION OF THE GENERAL EQUATION

Equations 3-14 and 3-15 are important, and their application is worth some study. Let us investigate two functions of engineering interest.

1. Given the experimental function

$$R = kX^b \tag{3-16}$$

with k and b constant, what is the standard deviation in the result R for a given standard deviation s_x in the variable measurement X? Applying Eq. 3-15,

$$\frac{dR}{dX} = bkX^{b-1} \tag{3-17}$$

and

$$s_c^2 = b^2 k^2 X^{2(b-1)}s_x^2 \tag{3-18}$$

It is convenient, when possible, to write such equations in terms of percent error. Dividing Eq. 3-18 by Eq. 3-16, and squaring, we obtain

$$\frac{s_r}{R} = b \frac{s_x}{X} \tag{3-19}$$

Example 3-2 The flow of water over a triangular weir is a function of the head of water to the 2.5 power. If the standard deviation in head reading is 3% of the absolute value of head reading, what is the percent error in flow, assuming that any other quantities are precisely measured?

Solution From Eq. 3-19,

$$\frac{s_r}{R} = 2.5 \times 0.03 = 0.075 \qquad \text{or} \qquad 7.5\%$$

Notice that, if an exponent is less than 1, the error in the result can actually be smaller than the error in the measured quantity.

2. Given the function

$$R = \frac{XY}{X + Y} \tag{3-20}$$

we see that

$$\left(\frac{\partial R}{\partial X}\right)_y = \frac{Y(X + Y) - XY}{(X + Y)^2} \qquad \text{and} \qquad \left(\frac{\partial R}{\partial Y}\right)_x = \frac{X(X + Y) - XY}{(X + Y)^2}$$

Then Eq. 3-15 gives (assuming, for example, that $w = p$ where p is the probable error)

$$p_r^2 = \frac{Y^4}{(X + Y)^4} p_x^2 + \frac{X^4}{(X + Y)^4} p_y^2$$

This result is usable, but it is usually better to obtain the percent error if doing so does not give a more complex function. Introducing Eq. 3-20 with both sides squared,

$$\frac{p_r^2}{R^2} = \frac{Y^4}{(X + Y)^4} \frac{(X + Y)^2}{X^2 Y^2} p_x^2 + \frac{X^4}{(X + Y)^4} \frac{(X + Y)^2}{X^2 Y^2} p_y^2$$

Performing the indicated algebraic operations, we obtain

$$\frac{p_r^2}{R^2} = \left(\frac{Y}{X+Y}\right)^2 \frac{p_x^2}{X^2} + \left(\frac{X}{X+Y}\right)^2 \frac{p_y^2}{Y^2} \qquad (3\text{-}21)$$

Example 3-3 The equation of the overall heat-transfer coefficient U for a system of two fluids separated by a wall of negligible thermal resistance is

$$U = \left(\frac{1}{h_1} + \frac{1}{h_2}\right)^{-1} = \frac{h_1 h_2}{h_1 + h_2}$$

where h_1 and h_2 are the individual coefficients of the two fluids. If h_1 is 15 Btu/h $\cdot$ °F $\cdot$ ft^2 with a probable error of 5%, and h_2 is 20 Btu/h $\cdot$ °F $\cdot$ ft^2 with a probable error of 3%, what will be the probable error in U?

Solution From Eq. 3-21,

$$\left(\frac{p_u}{U}\right)^2 = \left(\frac{20}{20+15}\right)^2 0.05^2 + \left(\frac{15}{20+15}\right)^2 0.03^2$$

$$\frac{p_u}{U} = 3.1\%$$

The propagated errors in certain basic functions are listed in Table 3-1. All can be found by the application of Eqs. 3-14 or 3-15. Many other such relationships can be computed following the methods outlined, and the

Table 3-1 A few equations of error

Function R	Error in result w_r
(a) $k(X + Y)$	$(w_x^2 + w_y^2)^{1/2}$
(b) kXY and	$R[(w_x/X)^2 + (w_y/Y)^2]^{1/2}$
(c) kX/Y	
(d) kX^b	bRw_x/X
(e) ke^x	Rw_x
(f) $k \ln X$	$Rw_x/(X \log X)$
(g) $k \sin X$	$Rw_x/\tan X$

reader is warned against too careless a use of the few equations given in this text. For example, the equation given in Table 3-1 for the function

$$R = ke^x$$

is entirely incorrect for the function

$$R = ke^{-x}$$

3-4 PLANNING EXPERIMENTS FROM AN ERROR ANALYSIS

Probably the most useful application of the ideas we have discussed in the previous sections occurs in the planning portion of the engineering experiment. Often the investigator will have a wide choice of instrument and test rig complexity and logically will wish to maintain a reasonable simplicity and expense level. One does not have to look very far in most industrial or government laboratories to see experiments in which instruments of all accuracy ranges are thrown indiscriminately into apparatus. Tens of thousands of dollars may be spent on the best possible instrumentation to obtain final data that appear once in a half-page graph, of which the "standard deviation of interpretation" may easily be 10% or more. The supervisor or test engineer in charge who permits such slipshod planning is costing the facility huge sums that might otherwise be available for new and pressing research. The following two examples show how error analysis can be useful in a planning situation.

> **Example 3-4** Figure 3-4 shows the instrumentation for a typical heat-exchange test. We wish to obtain the heat flow in or out of the water from $q = wC_p(T_h - T_c)$, where C_p is the specific heat, w is the flow rate, and $T_h - T_c$ is the temperature rise or fall in the fluid. The idea of the test is to obtain the overall heat-transfer coefficient UA from
>
> $$UA = \frac{q}{LMTD}$$
>
> where
>
> $$LMTD = \frac{\Delta T_{max} - \Delta T_{min}}{\ln (\Delta T_{max} - \Delta T_{min})}$$

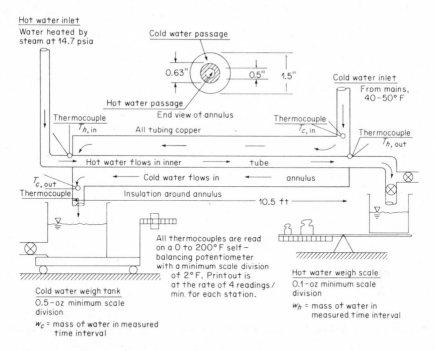

Figure 3-4 Typical heat-exchange test as described in Example 3-4.

and Fig. 3-5 shows the nomenclature of the log mean temperature difference (*LMTD*). A trial of six runs gives the following results:

Point no.	$T_{h,\text{in}}$ (°F)	ΔT_h (°F)	w_h (lb/min)	$T_{c,\text{in}}$ (°F)	ΔT_c (°F)	w_c (lb/min)	q_h (Btu/min)	q_c (Btu/min)
1-A	173	118	0.7	40	1.5	47.95	82.5	72
2-A	173	123	1.0	40	2.5	47.95	123	120
3-A	173	112	1.38	40	3.5	47.95	154.5	168
4-A	172	84.5	3.05	40	6.0	47.95	258	288
5-A	172	61.5	8.4	40	11.0	47.95	512	530
6-A	171	50.0	12.3	40	13.0	47.95	615	624

Discuss the instrumentation errors, their propagation, and the expected error in q and UA.

Solution The table shows a serious heat-balance error in the experiment, since from the first law of thermodynamics we expect the energy into the

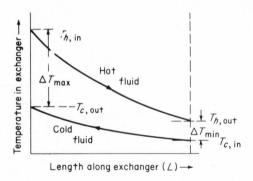

Figure 3-5 Showing the nomenclature of the quantity *LMTD* in Example 3-4.

cold fluid to equal the energy lost from the hot fluid. Only in run 2 is this ideal approximated. The hot water flow is noted to have a weigh scale with 0.1-oz increments. We can thus expect for a 1-min weighing period a worst-case error in w_h at the minimum condition of $0.1/(16 \times 0.7)$, or 0.9%. The worst-case, cold-water error with a minimum scale division of 0.5 oz is even smaller, $0.5/(16 \times 47.95)$, or not even 0.1%. However, if we consider that reading (or holding constant) temperatures to closer than $0.5°F$ is unlikely with standard lab instrumentation and controls, we can estimate the error in $\Delta T_c = (T_{c,\text{out}} - T_{c,\text{in}})$ from Eq. *a* in Table 3-1 as

$$w_{\Delta T} = (w_{tc,\text{out}}^2 + w_{tc,\text{in}}^2)^{1/2} = (0.5^2 + 0.5^2)^{1/2} = 0.707°F$$

Since the worst-case (smallest) ΔT_c is only $1.5°F$ we have a possible error of $0.707/1.5$, or 47%, in q. Even with the best case, run 6, where ΔT_c is $13°F$, the expected maximum error is $0.707/13$, or almost 6%.

Based on this analysis, we should use q_h in our calculations to find UA, since neither the hot flow nor the temperature errors should exceed 2%. However, we cannot discover from this data how much in error q_h might be.

We shall leave as an exercise the examination of the LMTD function's behavior with uncertainties of $0.707°F$ in the ΔT quantities, since this will affect the uncertainty of UA.

How should the test be improved? Clearly, the cold water should be run more slowly and its ΔT increased. Low-water flows should probably be measured for more than 1 min. Most important, the test should certainly be run with greater time between points to ensure that steady-state conditions exist on both the hot and cold sides.

Although all continuous functions can be differentiated, the results may be so complex as to make the error analysis far more difficult than final analysis of the data itself. We can, however, write Eq. 3-15 in *finite difference form*

$$w_r^2 = \left(\frac{\Delta R}{\Delta X_c}\right)_y^2 w_x^2 + \left(\frac{\Delta R}{\Delta Y_c}\right)_x^2 w_y^2 \qquad (3\text{-}22)$$

and find the numerical value of the derivatives by simply allowing ΔX to change by some small but finite amount and finding the resulting ΔR, thereby forming the finite derivatives in Eq. 3-22.

Each instrument or parameter error contributes its share of error to the result. A way of comparing the relative *influence* of each variable variation separately on the results is to define a *sensitivity ratio* (SR) from

$$SR = \left(\frac{\Delta R/R}{\Delta X/X}\right)_{y, z, \ldots} \qquad (3\text{-}23)$$

The sensitivity ratio is simply the ratio of the *percent change* in R, the result, divided by the percent change in the variable X that produces this change in R. A SR of less than 1.0 shows that the X variable does not propagate its error as strongly as its own variation, whereas a SR of greater than 1.0 shows that the X change has a greater percentage effect on R than on itself.

Example 3-5 A student wishes to determine the *virtual mass* M_v of the human body in a "tuck" or "cannonball" position by utilizing the equation[*]

$$M_v = \frac{AC_d \, \rho \, D_{\max}}{\ln \, [1.0 + AC_d \, \rho \, V_0^2/2F_b]} \qquad (3\text{-}24)$$

The virtual mass is made up of two parts, the *actual mass M* of the body, which in this case is a 72.56-kg (160 lb) student, and the *added mass* M_a, which is produced when a body is accelerated or decelerated in a fluid. Some of the fluid accelerates with the body and makes it appear to be more massive and thus respond to a given force with lower acceleration. The cannonball position was chosen because in it the body approximates a sphere and the added mass for a sphere has been found to be equal to just one-half of the water displaced. Thus, in this case, since the human body displaces almost exactly its own weight in water, we predict a virtual mass of 72.56 + (72.56/2) kg, or 108.84 kg.

[*]See Schenck, 1970, pp. 21–42.

The value V_0^2 is found from the usual physics equations for falling bodies $V_0^2 = 2gH$ and represents the velocity of entrance to the water. In this case, the center of the tucked student was measured to be 1.35 m from the water surface when the fall started, giving, with $g = 9.8$ m/s^2, a V_0^2 value of 26.46 m^2/s^2.

Preliminary trials showed that the natural buoyancy of the subject was not enough to ensure not hitting the bottom of the pool, so a Neoprene wet suit was worn and the buoyant force of the subject F_b was measured under water with a spring scale at the bottom of the pool, giving a buoyant force of 70.06 N (15.75 lb$_f$). The value of D_{max}, the depth reached by the center of the tucked subject, was observed by a diver against a long, vertical scale.

The biggest problem with the experiment is the value AC_d. This is known as the *drag area* and is equal to the product of the *drag coefficient* C_d of the body times its frontal area A. The only reference available[*] gave a value of 0.279 m^2 (3 ft^2), but this had been obtained in a wind tunnel test on a stationary, tucked human subject facing the wind. We have both precision and accuracy problems with AC_d. A human falling into the water is not duplicating the wind tunnel test very closely, so that the actual or true AC_d value is likely to be different to some degree from that found in the tunnel test. In addition, since the attitude of the subject striking the water is bound to change a little on each trial, the actual C_d may change from trial to trial. How should we go about analyzing this experiment?

Solution It would be hard to find an experimental function more suited to finite-difference methods than this one. If Eq. 3-24 is partial differentiated in terms of the several variables with error to form Eq. 3-15, the resulting mathematical expression is almost unmanageable. Instead, let us apply the idea of a finite difference, or approximate calculation. We shall do this in detail with AC_d and then simply give the results for the other variables.

We must first use the given parameters to obtain a predicted D_{max}, using the "theoretical" or anticipated value of M_v, 108.84 kg. Thus, with ρ for water equal to 1000 kg/m^3,

$$D_{max} = \frac{108.84}{0.279 \times 1000} \ln \left(1 + \frac{0.279 \times 1000 \times 26.46}{2 \times 70.06}\right) = 1.544 \text{ m}$$

Now let us choose a small but finite variation in AC_d to form the derivative, $\Delta M_v / \Delta A C_d$. We increase AC_d by an arbitrary 5% (new value

[*]Hoehner, 1965.

0.279 + 0.0139 or 0.2929 m^2) and solve for the new M_v, holding all other parameters at their fixed or nominal values:

$$M_v = \frac{1.544 \times 0.2929 \times 1000}{\ln [1.0 + (0.2929 \times 1000 \times 26.46)/(2 \times 70.06)]} = 112.22 \text{ kg}$$

Then

$$\frac{\Delta M_v}{\Delta A C_d} = \frac{112.22 - 108.84}{0.2929 - 0.279} = 243.17 \text{ kg/m}^2$$

which we shall use later in Eq. 3-22.

Let us find the sensitivity ratio for this variable. From Eq. 3-23,

$$SR = \frac{(112.22 - 108.84)/108.84}{(0.2929 - 0.279)/0.279} = 0.622$$

This is a favorable result. The hard-to-estimate AC_d error will propagate only about two-thirds of its own percentage error into the dependent result.

The same approach was used to obtain the derivatives and SR values for the other variables and to form an *error table*. In each case, the independent variable was increased by 5% and the change in M_v calculated.

Note that the sensitivity ratio (col. 4) is highest for D_{max}. Fortunately, the student found this to be one of the measurements having the smallest percent uncertainty. F_b has the low SR of 0.193 and so can have a high random error without seriously prejudicing the test. Notice that even if we have no idea about the error magnitude (col. 5), the SR tells us which measurements need the most care and may suggest where error sources lie.

We can now find the predicted uncertainty in M_v from the finite-difference form of the general error equation, Eq. 3-22, but with four terms, using cols. 3 and 5 from the error table:

$$w_{mv}^2 = 243.2^2 \, 0.0279^2 + (-1.27)^2 \, 2.911^2 + 0.3^2 \, 2.25^2 + 69.7^2 \, 0.075^2$$

$$= 46.04 + 13.67 + 0.46 + 27.33 = 87.5 \text{ kg}^2$$

Notice that the assumed C_dA error gives the largest contribution to w_{mv}, with D_{max} being next in importance. The F_b error could be much larger and still not affect the error in M_v.

Then $w_{mv} = 9.35$ kg, and we expect that the M_v data will show a

Error table for Example 3-5

Quantity (1)	Nominal value (2)	$\dfrac{\Delta R}{\Delta X}$ (3)	Sensitivity ratio (SR) (4)	Estimated error (5)	Remarks (6)
AC_d	0.279 m²	+243.2 kg/m²	0.633	0.0279 m²	Pure guess; take 10% of nominal as maximum error
V_0^2	26.46 m²/s²	−1.27 kg/m² · s²	0.309	2.911 m²/s²	Assumed a maximum error in height of 0.15 m or 11%
F_b	70.06 N	+0.30 kg/N	0.193	2.25 N	Subject "bobbed" during F_b measurement ±4.5 N, so we took maximum uncertainty as half this amount
D_{max}	1.544 m	69.7 kg/m	1.00	0.075 m	"Since subject dropped next to measuring stick, sighting was easy and within ±3 in"
ρ	1000 kg/m³				No error assumed
M (subject mass)	72.56 kg				No error assumed

maximum percent error of 9.35/108.84, or 8.6%. This is certainly a very reasonable error projection for finding the virtual mass of any complex object, particularly a human being, and we conclude that we should press forward.

Obviously, if the only point in error analysis were to get the test underway, we would now leave this example. But at least half its benefit lies in the way such error studies allow one to analyze the results. Let us look at some actual data and interpret them in terms of the error table and other ideas of this chapter. Six trials produced the following:

D_{max} (m)	1.851	1.851	1.775	1.751	1.781	1.979
M_v (Eq. 3-24) (kg)	131.44	131.44	126.1	123.4	127.5	134.4

The mean is 129.04 kg and the SD is 5.85 kg. Twice this, assuming that this encloses "most" of the errors, is 11.7 kg, and the maximum percentage uncertainty is thus 11.7/129.04, or 9.1%, compared with the predicted 8.6%. Again, remember that we cannot discover by repeated runs any accuracy or bias error in AC_d.

The experimental added mass M_a is M_v minus the subject mass, or $129.04 - 72.56 = 56.48$ kg. All of the uncertainty in M_v is also in M_a, since M is known exactly. Thus the maximum uncertainty in the added mass M_a is 11.7/56.48, or 20.7%. Now, we noted at the start of this exercise that the theoretical and accepted added mass for a sphere was half the displaced water, or 36.3 kg in this case, which is lower than can be explained by the random errors in the test. We might wonder, however, how much bias would be needed in our assumed constant AC_d to correct the experimental added mass to the theoretical value. This would be a correction of $56.48 - 36.3$, or 20.18 kg, which is 15.6% lower than the experimental mean value of M_v (129.04 kg). The error table shows us that the SR of AC_d is 0.622 and thus it must change 0.156/0.622, or 25.1%. Thus the tabulated value of AC_d must be about 25% smaller to correct, by itself, our M_v to the theoretical value. It is much more reasonable simply to assume that the M_v of a tucked human body is somewhat higher than that for the sphere. Certainly, the feet, elbows, and head will entrain additional water. Also, the body probably has some spin or is given some at entrance, and this will also entrain water. Lacking further data, it is reasonable to accept the student's M_v values with the noted errors.

3-5 UNCERTAINTY PROPAGATION
USING CHARTS AND CURVES

So far in this chapter we have assumed that the function relating the measurements X, Y, Z to the result R was known, and we then applied the special or general methods of Table 3-1 or Eqs. 3-14 or 3-15 to this known function. In many engineering situations, instrument readings are processed using functional relationships in the form of curves, function scales, charts, nomograms, or tables. To utilize such pictorial, graphic, or tabular functions we must again apply a *finite-difference* approach. We might, for example, use an instrument yielding a reading X, which we estimate to have an uncertainty w_x and wish to use with a curve of R versus X to obtain the result R. The general equation for this situation from Eq. 3-15 is

$$w_c = \frac{dR}{dX_c} w_x \qquad (3\text{-}25)$$

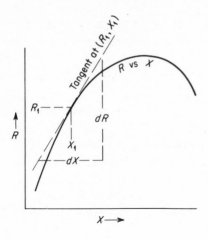

Figure 3-6 A sketch showing one method of finding a derivative graphically for use in error propagation formulas.

If Fig. 3-6 represents the R-versus-X function, we can obtain the derivative dR/dX graphically by drawing the tangent at X and measuring its slope. Equation 3-25 is then easily applied if w_x can be estimated.

In some situations, the use of a smoothed curve introduces uncertainty or error even if the instrument readings are perfectly precise. The universally used Mollier chart for steam and its associated tables are considered to have about a maximum uncertainty of 1 Btu/lb. Thus, if we enter these tables with temperatures or pressures having their own uncertainty, we must consider all these various deviations.

Example 3-6 A steam calorimeter gives readings for pressure and temperature of the throttled steam. We note a pressure of 15 psia and a temperature of 240°F. A possible maximum uncertainty in pressure is 0.2 psia and in temperature, 5°F, this latter figure being due to the impossibility of perfectly insulating the calorimeter. What will be the maximum uncertainty in enthalpy as read from tables?

Solution The steam tables[*] give the following tabulation of enthalpy in the region of interest (in Btu/lb):

Absolute pressure (psia)	Temperature (°F)		
	220	240	260
14	1154.6	1164.4	1174.0
15	1154.3	1164.1	1173.7
16	1153.8	1163.7	1173.4

[*]Keenan and Keyes, 1936.

The general equation for uncertainty in h as a function of uncertainties in T and P is simply Eq. 3-22 in the form

$$w_h^2 = \left(\frac{\Delta h}{\Delta T}\right)_P^2 w_t^2 + \left(\frac{\Delta h}{\Delta P}\right)_T^2 w_p^2$$

We wish first to evaluate $(\Delta h/\Delta T)_P$ around the h value at 15 psia and $240°F$. From the small table we see that

$$\left(\frac{\Delta h}{\Delta T}\right)_P = \frac{1173.7 - 1154.3}{40}$$

or $(\Delta h/\Delta T)_P \approx 0.48$. Making a similar finite-difference approximation, $(\Delta h/\Delta P)_T = 0.35$. Notice that we do not keep track of signs since we shall square both quantities. The application of Eq. 3-22 can now be made as follows:

$$w_h^2 = 0.48^2 5^2 + 0.35^2 0.2^2 = 5.8 + 0.005 = 5.8^+$$

Notice that the pressure uncertainty is insignificant in contributing to the final uncertainty in enthalpy. Then w_h is ± 2.4 Btu/lb with the uncertainties as given, assuming that the tables are exactly correct. However, we noted that there may be an uncertainty of as much as 1 Btu/lb in the table entries. We would certainly not expect the values in the small portion of the table considered to vary this much from one another, but it might be that this part of the table would be off by as much as 1 Btu/lb overall. Then the enthalpy has two possible uncertainty sources: the basic "smoothing uncertainty" in the tables themselves, having a magnitude of 1 Btu/lb or less; or the propagated uncertainty calculated as 2.4 Btu/lb. It is not obvious how these uncertainties should be best combined. Actually, we would probably be content with the two separate figures. We see that our calorimeter has not yet reached the potential limit of accuracy set by tabular uncertainties and that we should attempt to reduce the temperature uncertainty. If this can be brought to $\pm 2°F$, the net error in h due to T and P uncertainties would drop to 0.76 Btu/lb, somewhat less than the maximum tabular uncertainty. It is doubtful whether any further improvement in sampling precision would be justified.

Example 3-7 In Example 2-5 we discussed the error distribution of 11 student measurements of $F_{1 \rightarrow 2}$ using an angle-factor integrator. The student measurements showed a bias error of about 0.6%/13.87%, or about 4.3% below the true value. The maximum random error percentage, based on a 2 SD interval, was about 7%. Thus the worst that a single reading might be off was about 9% too low. The system solved by the students can also be solved using the equation

$$F_{1 \to 2} = 1.75F_a - 0.75F_b$$

where F_a and F_b are simpler angle factors that have been charted in texts and reference works on heat transfer. Assuming that we can read the two text graphs to about the same precision, how close must we read (or how close must the printed graph be to the true value) so that the uncertainty in $F_{1 \to 2}$ will not exceed 9%?

Solution Nine percent of the true value of 13.87% is 0.0125. Then, if $w_{fa} = w_{fb}$, Eq. 3-15 gives $0.0125^2 = 1.75^2 w_{fa}^2 + 0.75^2 w_{fb}^2$, and w_f is 0.0066, or 0.66%. The maximum value on any angle-factor tabulation is unity, and we would have to read each graph to within two-thirds of a percent. Now there is no way that the average, half-page text graph can even be drawn, much less read, to this precision (two-thirds of a percent of full scale), and it was for this reason that the student results were checked by a computer solution. Thus we can make "experimental" errors when reading graphed data, just as much as when we read an instrument. In this interesting example, the experimental device, provided that its limitations are understood, is probably more accurate than graphed solutions taken directly from reference books. Further, this cautionary example may deter those who regard what is on the printed page as absolutely correct and without uncertainty.

3-6 LINEAR PROPAGATION
AND UNCERTAIN CONSTANTS

The propagation formulas used herein are well accepted by statisticians and others in measurement specialties. It is still possible, however, to find books and articles in engineering and physics[*] that do not utilize the basic square-propagation formula as shown in Eqs. 3-14 and 3-15. Since it is little more work

---

[*]The interested reader is referred to the following series of articles and notes that discuss these methods: Birge, 1939; Fleisher and Olsen, 1950; Hanau, 1951; and Rowles, 1959.

to square the percentages, such an approach is now considered old-fashioned and will not give so valid an estimate of the random error or uncertainty in a result as will the application of the methods described here.

The reader may occasionally be faced with a fixed measurement or constant that will enter all computations in the same way. Examples of such a fixed quantity are lever-arm lengths, resistances in Wheatstone bridges, tare or zero readings, physical and mechanical constants such as a specific heat or spring constant, and "natural" constants such as the velocity of light, gravitational acceleration, and so forth. Such quantities are not actually random variables or susceptible to a changing precision uncertainty during a test. Once we select a lever-arm length, for instance, we are committed to that single value over a number of different runs. Nevertheless, we shall consider such constants to be susceptible to random fluctuation and thus to have an estimated uncertainty with a distribution of values about the correct one. This is a perfectly reasonable approach and simply means that during a test we shall not replicate the readings of such constants. Obviously, if it should happen that the estimated uncertainty in such an unchecked quantity is too high, we shall replicate it if possible. The following example may make this point more clear.

Example 3-8 Figure 3-7 shows a simple Wheatstone bridge circuit to be constructed of standard resistances. The unknown resistance R_u is a variable sensing element in a test apparatus, and we wish the relay to open when null current occurs between the bridge arms. This will occur when

$$R_u = \frac{R_x R_y}{R_x}$$

Let us assume that the known resistances will be drawn from catalog stocks. What must their tolerance be, in percent, if R_u is not to deviate from a design value by more than 5%?

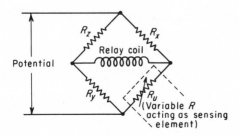

Figure 3-7 The Wheatstone bridge null-sensing circuit discussed in Example 3-8.

Solution Here is a case in which we actually have no random variable at all. Once we select the three resistances, the apparatus will always have a null point at a fixed R_u. Yet we can visualize that each resistor comes from a population having a symmetrical distribution falling within some, yet unknown, limits. Let us use Eq. 3-11, where the ws now represent some maximum deviation (enclosing perhaps 19 in 20 or possibly 99 in 100). Then $0.05^2 = 3(w_R/R)^2$, and $w_R/R = 0.029$, assuming that we wish all three resistors to have the same tolerance, and this would undoubtedly serve in this application. Thus, even in a situation in which there is no real random variable at all, it is possible to imagine the fixed values to have a deviation distribution and proceed accordingly.

3-7 PROPAGATION OF NONNORMALLY DISTRIBUTED ERRORS

In the derivation of Eq. 3-14, there is no requirement that the error distributions being combined be normal. Since Eq. 2-21 is simply a definition of s, and since the summation of the squared error terms comes directly from a simple algebraic manipulation, Eq. 3-14 will be entirely correct even if one or more of the combined distributions are flat, peaked, triangular, or of the "roulette wheel" type. The only requirement imposed on Eq. 3-14 by its derivation is that no error distribution be excessively skewed such that the $\Sigma\, xy$ terms do not sum to zero as n becomes large.

Equation 3-15 (and Eqs. 3-11, 3-21, and those in Table 3-1) involves an additional assumption: that the relation between w (any given uncertainty interval) or p (the probable error) and the standard deviation s is the same for all error distributions. Mathematically, the requirement may be stated as

$$w_x = k_1 s_x \qquad w_y = k_2 s_y \qquad \text{etc.}$$

but

$$k_1 = k_2 = \ldots k_n \tag{3-26}$$

These proportionality constants (the ks) will be equal only if the various error distributions are all of the same type, the normal distributions usually being assumed.

Suppose that we wish to propagate the error distributions shown in Fig. 2-15 in a sum-type function as

$$R = C(A + B)$$

where C is a fitting constant without error. This leads (see Table 3-1) to

$$s_r = (s_a^2 + s_b^2)^{1/2} \tag{3-27}$$

The A distribution was found to be approximately normal; so

$$p_a = 0.675s_a \tag{3-28}$$

The B distribution had an s_b of 0.58 in and a p_b of 0.5 in; so

$$p_b = 0.86s_b \tag{3-29}$$

Clearly, the requirements just stated for writing Eq. 3-24 in the form found in Table 3-1,

$$p_r = (p_a^2 + p_b^2)^{1/2} \tag{3-30}$$

are *not* met. The two constants noted in Eq. 3-26 are not the same. In addition, we have no easy way of knowing just what the relationship between s_r and p_r will be.

Now suppose that we start with known probable errors for the A and B distributions, change them to standard deviations using Eqs. 3-28 and 3-29, and combine these s values in Eq. 3-27 to obtain the standard deviation of the result. If we know that distribution B is of the roulette wheel type, we can obtain the correct s_b of 0.58 in from Eq. 3-29. With an s_a of 0.749 in, Eq. 3-27 gives the *true* s_r of 0.945 in. Note, however, that although we know the standard deviation of the error distribution of the result, we do not know what percentage of the total this plus-or-minus error may enclose. Remember that $\pm s_r$ encloses only about 68% of all the readings *if all the distributions involved are normal.*

Now imagine that we incorrectly assumed that the B distribution was of the normal type. We would now take the known p_b of 0.5 in and use the normal curve relationship, Eq. 3-28, to obtain an erroneous s_b of 0.74 in. Now Eq. 3-24 gives the *wrong* s_r of 1.02 in compared with the correct value of 0.945 in. This modest error of 10% in s_r would be considered negligible in most computations of this type.

It is true that the more usual analysis would be to go in the opposite direction, from s to p (or w). Still, it should be evident from this example that the use of Eq. 3-15 or Table 3-1 equations with nonnormal error distributions is not apt to produce serious mistakes in overall estimates. *Vastly more bad testing will result from the simple failure to apply Eq. 3-15 than will ever occur from using it with doubtful error estimates.*

When the test is a complex or sophisticated one, involving error-prone instruments, doubtful tabular data, and perhaps skewed or pathological error distributions, and when financial or humanitarian considerations require the most exhaustive sort of analysis of the test performance, the engineer should usually resort to a computer simulation of the complete instrument-plus-test-rig system. In this sort of analysis, the complete experiment is actually run on a computer over and over again with proper error distributions being sampled to provide realistic readings for each simulated run.

3-8 INSTRUCTIONS FOR USING CHAPTERS 2 AND 3

To plan the instrument-accuracy parts of an engineering test properly, you will need the following information:

1. *Accuracy* data on the instruments, which may involve *calibration* (Sec. 2-2), the preparation of *calibration curves* (Fig. 2-1), or some assurance that scales and previous calibration data are still all right.
2. *Precision* data on the instruments, which may be detected by *replicating*, or returning to, standard inputs and observing how the output varies (Sec. 2-3). Such precision error may result from *extraneous environmental effects* (Example 2-1), *hysteresis* (Example 2-2), or a variety of uncataloged and small variations between each replication (Fig. 2-3).
3. The *experimental functions*, such as efficiency, Reynolds number, *LMTD,* and so on, which will be computed from the raw or calibrated instrument readings (Secs. 3-1, 3-2, 3-3).

Accuracy error is then eliminated by correcting the raw instrument output using the calibration curves or equations. Precision error is studied as a random variable. Unless tests for normality show differently (Sec. 2-8), we assume the error distribution follows the theoretical, normal law equations described in Secs. 2-4 and 2-5, and adjusted for use with finite samples of data in Secs. 2-6 and 2-7. The two most essential statistics of the sample are its *mean* (Sec. 2-6) and its *standard deviation* (SD) (Sec. 2-7). These, in conjunction with Table 2-2, allow us to predict chance limits on finding a given error or excluding a given error (Example 2-4). Other statistics that have use in the study of normally deviating errors are the *probable error* (Eq. 2-24) and the *standard deviation of the mean* (Eq. 2-25). If we cannot replicate inputs or otherwise experimentally obtain the precision error of an instrument, we should *estimate* it (Sec. 2-9) if possible. Normally deviating data may also be studied in its own right, that is, not as part of an instrumented experiment (Sec. 2-10), using the methods of Chap. 2.

Knowing the SDs of the test instruments and the functions their readings will enter, we next estimate the *propagated precision error* to be expected, using the methods of Chap. 3. This can be done using calculus (Secs. 3-2 and 3-3) or finite differences (Secs. 3-4 and 3-5). Often we can study the *relative* error effects of several instruments on the result, even if we do not know the precision errors themselves, using a *sensitivity ratio* (Eq. 3-23), and compiling an *error table* (Example 3-5) that reveals those readings in a test having the greatest (and least) effects on the precision of the result. Charts, tables, and graphs can be as large a source of uncertainty as the instruments (Sec. 3-5), and there may be *experimental constants* (tare weight, lever arm, drag coefficient) that we do not remeasure at each data point but that we assume have a distribution of uncertainty or precision error like those readings taken each time (Sec. 3-7).

The entire study of instrument error is a *feedback system*. The estimation of precision indexes and the assembling of these to the estimation of error in the final result may call up changes in the instruments, the way of obtaining one or another theoretical values (Example 3-7), or the way the test is operated (Example 3-4). The test itself may suddenly require a new start on the matter, when output is so random as to be useless. Sometimes it may even be necessary to seek a new mathematical approach to the test or the measurements, or at least a new set of test parameters. Chapter 4 will assist in this question.

PROBLEMS

3-1 Prove that, when $R = X/Y$, the analytic method outlined in Sec. 3-1 (not the general method) gives the error equation, Eq. 3-8.

3-2 We wish to construct a 50-Ω resistor by combining two 100-Ω resistors in parallel. If the 50-Ω unit is to have an error no greater than 1%, what accuracy limits (in percent) must we place on the two 100-Ω units?

3-3 Obtain one or more of the following items from Table 3-1 using the general equation for probable error (Eq. 3-15): *a, d, e, f, g.*

3-4 A vessel navigation system using an inertial mass of 0.1 kg but with a SD of 0.001 kg around this figure will travel in regions where g (the gravitational acceleration) will vary with a SD of 0.011 m/s² with a mean of 9.800 m/s². If $F = Mg$ and the force in newtons is the sensed result, what is the percent error based on a ±95% range in F due to the two variations? We can correct for g by knowing our location. Is this worthwhile?

3-5 The height h of a mercury column in a mercury barometer is found from $h = P_a/(\rho g)$, where P_a is the absolute pressure to be measured by the instrument. Because of wetting and meniscus errors, we assume a reading of h as a nominal 0.76 m, or 760 mm, with a SD of 1 mm (0.01 cm). Density, ρ, is a nominal 13.6 g/cm³ and a SD of 0.1 g/cm³ due to temperature effects. The value of g varies as described in Prob. 3-4. What is the total uncertainty as measured by its SD in P_a, and which of the three error sources should be dealt with first?

3-6 The Carrier equation gives the actual vapor pressure P_v in terms of the saturated vapor pressure at the wet-bulb temperature P_{sat}, the barometric pressure P_b, the dry-bulb temperature T_{db}, and the wet-bulb temperature T_{wb} as follows:

$$P_v = P_{sat} - \frac{(P_b - P_{sat})\,(T_{db} - T_{wb})}{2800 - 1.3T_{wb}}$$

With P_{sat} of 0.178 psia, T_{wb} of 50°F, T_{db} of 70°F, P_b of 14.68 psia, what will be the percentage uncertainty in P_v if all the quantities are error-free except P_{sat} and T_{wb}, which may have uncertainties of ±6%?

3-7 The orifice coefficient C for a flat-plate orifice is given by

$$C = q\,\frac{[1 - (A_2/A_1)^2]^{1/2}}{A_2\,(2gh)^{1/2}}$$

q, the volume rate of flow, equals 20 ft³/min with no error; A_1 is the pipe area, 10 in²; A_2 is the orifice area, 5 in²; and each has a probable error of 0.1 in²; g is 32.2 ft/s²; and h is the pressure change across the orifice equaling 32 in of the flowing fluid with p_h of 0.1 in. Find the error in C, and decide which of the measurements should be increased in accuracy for greatest improvement in the precision of C.

3-8 A hinged-end colume with compression load P of 1000 ± 50 lb load, eccentricity e of 0.01 ± 0.001 in, deflection δ to be found, moment of inertia I of 1 in⁴, modulus of elasticity E of $10 \times 10^6 \pm 0.5 \times 10^6$ psi, and length L of 12 in follows the equation

$$\delta = e\left\{\sec\left[\left(\frac{P}{EI}\right)^{1/2}\frac{L}{2}\right] - 1\right\}$$

What will be the error in deflection?

3-9 *Young's modulus* E is defined as the *stress*, in newtons per square meter (N/m²), of a specimen divided by the *strain*, in meters of stretch divided by meters of length. Thus, $E = (F/A)/(\Delta L/L)$. If F/A is read from a testing machine dial with a maximum (2 SD) uncertainty of 0.6%, L is read with a micrometer to 0.05%, and ΔL is read by an extensometer to 5%, what is the error in E?

3-10 A student proposes that the *tidal volume* of a subject, the maximum amount of air one can take into one's lungs, be measured by placing the subject in water and testing the change in buoyancy between full inhale and full exhale. The subject will be given some initial positive buoyancy so that the change is always in the same direction. The volume of the lung $V_1 = (F_f - F_e)/(\rho g)$, where F_f is the full-lung upward force measured on a spring scale, F_e is the empty-lung upward force, and ρ is the density of the water. ρ will have a nominal value of 0.9997 g/cm³ and a SD of about 0.0001 g/cm³ due to temperature changes, but we can reduce this if we make temperature measurements during the test. If the lung volume V_1 is a nominal 0.006 m³ and the readings of F_f and F_e are made with a scale having a least count of 0.14 N (0.5 oz) and g is 9.8 m/s² with no error, what will be the actual and percent error in V_1? Is it worthwhile bothering with the water-density changes?

3-11 We wish to measure the refractive index N of a transparent substance by measuring the angle of the incident beam θ_i and of the refracted beam θ_r and solving the equation

$$N = \frac{\sin \theta_i}{\sin \theta_r}$$

If the index is in the neighborhood of 1.5 and the incident beam can take any angle θ_i between 0 and 90° and either angle has an uncertainty at all angles of ±0.2°, at what angle of θ_i should we run the test?

3-12 The log mean temperature difference is defined by

$$LMTD = \frac{\Delta T_1 - \Delta T_2}{\ln (\Delta T_1 / \Delta T_2)}$$

If $\Delta T_1 = 2 \ \Delta T_2$ and each has the same percent error, what must be this error for $LMTD$ to have a percent error of no more than 5?

3-13 The true pressure differential in a test system is to be found from a two-fluid manometer in the range of 0 to 1 in of water. Suppose that h is the indicated difference in liquid heights with an error of 4%, SG_1 is the specific gravity of the lighter fluid and SG_2 the specific gravity of the heavier fluid, and R is the ratio of tube area to reservoir area with a value of 0.10. An instrument is available to measure liquid specific gravities to within ±5%, and the following liquids are available: SG-A 0.86, SG-B 1.0, SG-C 1.08, SG-D 1.56, SG-E 2.2, SG-F 3.0. If the equation of the instrument is $\Delta P = h(SG_2 - SG_1 + R \times SG_1)$, select the two fluids that will give the most accurate determination of ΔP in the range desired.

3-14 The equation expressing the variation of resistance R with temperature T is $R = R_0 (1 + \alpha T)$, where R_0 and α are constants to be determined by test. If R is 10.3 Ω at 50°F and 11.7 Ω at 150°F and p_r is ±0.1 Ω and p_t is ±1°F, what is the error in R_0 and in α?

3-15 The length of a rod, L_0, is measured and then it is heated from the temperature at which L_0 was obtained to a new and higher temperature, for a rise of ΔT. The *coefficient of linear expansion* γ is measured from $L = L_0 (1 + \gamma \Delta T)$. If we have a substance in the range of 20×10^{-6} (1/°C), ΔT cannot go much more than 300°C but can be accurately measured, and we cannot measure the length L_0 or L to better than 0.5 mm, what must L_0 be to obtain γ to within ±5%?

3-16 The power developed in a motor is equal to $I^2 R$, where I is measured by a clamp-on ammeter and R with an ohmmeter. R is 0.2 Ω with a 5% maximum uncertainty and I is found to vary about ±1 A depending on what part of the wire we clamp over. Careful and independent measurement of line power loss gives 331 W taken when the current is measured to be 40 A. If six more readings give a mean of 331 W and a SD of ±14 W, does this seem in keeping with the expected uncertainties in R and I, with regard to both the variation and the location of the mean?

3-17 The light intensity I_d reaching a depth D in the ocean compared to its surface intensity I_0 is related to the *absorption coefficient* K by $I/I_0 = e^{-KD}$. I/I_0 is in the range 0.25 with a SD of 0.02 (so that the error is 0.02/0.25, or 8%). What will be the percent error in K if D has a nominal value of 10 m with a SD of measurement variation of 0.05 m? Which of the two measurements, I/I_0 or D, should we concentrate on improving if we wish to get a better estimate of K?

3-18 We wish to obtain the compressibility C_0 of scuba wet-suit material by measuring the increase in pressure ΔP, and noting the percent thinning, $\Delta L/L$ produced by the pressure increase, where $C_0 = (\Delta L/L)/\Delta P$. If L is 8 mm with a SD of ±0.5 mm, ΔP is 2×10^5 Pa, with a SD of 2%, and $\Delta L/L$ for this pressure is a nominal 0.5, predict the 95% error limits on C_0. What will be the percent error in C_0?

3-19 The figure below shows the schematic of a Stormer-type viscosimeter used to obtain the viscosity of oils, glycerines, and so on. Viscosity μ is found from

$$\mu = \frac{\Delta\theta}{1/\Delta m} \left[\frac{(b^2 - a^2)\, k^2 g}{4\pi a^2 b^2 S(l + e)} \right]$$

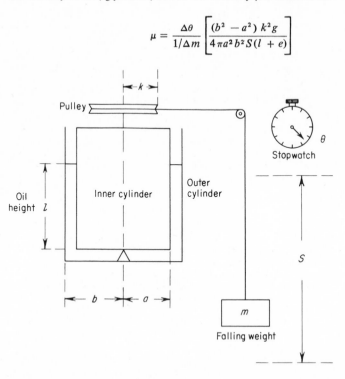

The quantities a, b, s, k, and l are measured with slides or calipers and we estimate a maximum possible uncertainty of about 0.03 in in each; e is given as 2.4 in exactly; Δm is read from "standard" masses and has no error, but the $\Delta\theta$ error is found by experiment to produce a maximum (2 SD) percent error of 7% in the $\Delta\theta/(1/\Delta m)$ slope determination. The error table gives the values and errors of the several measured quantities:

	Quantity						
	a	b	k	s	l	e	$\Delta\theta/(1/\Delta m)$
Magnitude	1.97 in	2.38 in	0.61 in	22.5 in	3.0 in	2.4 in	24.5 g · s
Max. uncertainty	0.03 in	0.03 in	0.03 in	0.03 in	0.03 in	0	7%

$$g = 386 \text{ in/s}^2 \text{ with no error}$$

Complete the error table by finding the finite derivatives and the sensitivity ratios for the quantities having error. What is the percent error in viscosity at this condition? Which measurement should be improved first?

3-20 One way of obtaining Young's modulus E (see Prob. 3-9) is to use the relationship between a sound wave velocity u, the density of the material ρ, and E, where $u = (E/\rho)^{1/2}$.

We obtain an acoustic velocity in the material of 1250 km/s with a material ρ of 9 kg/m³, but this value has an error of 0.5% because the material is not exactly homogeneous. The E we find after several trials disagrees with that found from testing machines by about 6%. What would the percent error in the sound velocity measurement have to be to give this error?

3-21 A so-called guarded hot plate consists of two heating plates so arranged that they are close together and at the same temperature. This means that the sides of the plates facing each other lose no heat, and all the heat into the plate q goes out one side. Then $q = HA(T_p - T_s)$, where H is the heat-transfer coefficient, A is the plate area, q is the measured electrical heating load on the plate, and the plate and surrounding air temperatures are T_p and T_s. The values of T_p (94°C) and T_s (14°C) are estimates with error since neither the plate nor the air is at constant temperature. If A is 0.093 m² and without error, q is 0.406 × 10⁶ J/h with a maximum error of ±6% due to fringing and heat loss at the edges of the plate, how low must the maximum error in the two temperatures be to obtain H with a maximum error of no more than 10%?

3-22 Taking the same uncertainties given in Example 3-6 for a steam measurement, estimate the propagated uncertainty in h for the region P of 500 psia and T of 1000°F.

3-23 What will be the percent error in viscosity, specific heat, thermal conductivity, and density of water at 100°F if the temperature is uncertain by 2°F? Repeat for air at 14.7 psia. Use either curves of properties or tabulations of properties from some standard handbook.

3-24 In a test of radiant heating panels q, the energy input to the panel is measured at 1000 Btu/h · ft² with no error assumed. The panel is at 570°R and radiates to a constant temperature sink at 490°R. Assuming that the absolute error in temperature (in degrees Rankine) is the same for either temperature measurement, how large can it become if the maximum error in emissivity e of the panel is to be less than ±10%? The equation is

$$q = FA\,\sigma\alpha\,e(T^2_{panel} - I^4_{sink})$$

where $FA\sigma\alpha$ are known constants without error and together equal 3.45 × 10⁻⁸ Btu/h · ft² · °R⁴. Note that e is a dimensionless ratio that lies between zero and 1.0.

3-25 The mean area of a seepage-flow system is given in terms of the entrance and exit areas A_1 and A_2 by

$$A_{mean} = \frac{A_1 - A_2}{\ln A_1 - \ln A_2}$$

A_1 is twice A_2, and the estimation error in each is 7%. What is the propagated error in A_{mean} due to these errors?

3-26 An experimental apparatus measures the ratio of the surface tension at $T(S)$ to the surface tension at 0°C (S_0) and the ratio of the surface film temperature T to the critical temperature of the fluid T^*. Such data are related by the equation

$$\frac{S}{S_0} = \left(1 - \frac{T}{T^*}\right)^n$$

and the purpose of the test is to find the important dimensionless parameter n. The maximum error in S/S_0 is 0.05 and in T/T^* is 0.03. What will be the maximum error in n in the region where S/S_0 and T/T^* are both equal to 0.5? Decide whether the test is a practical one and how it can be improved.

3-27 The rate of growth of a crystal is assumed to obey the law

$$T = Ce^{m/m_0}$$

where T is the time of growth, C is a kind of time constant based on the concentration of the solute with units in hours, and m/m_0 is the percent change in mass of the crystal. If the error in m/m_0 is 1% with m/m_0 equal to 2.0 and C is 0.3 h with an error of 0.05 h, what is the expected error in T?

3-28 The flow over a contracted weir W as a function of the measured head on the weir H is given by

$$W = 3.33(B + 0.2H)H^{3/2}$$

where B is the width of the weir and assumed known exactly. Obtain an equation for the error in W as a function of the error in H, with no other errors present. Using this formula, decide which of these two weirs (which pass the same flow) would be the better design: Weir X has a B of 1.4 ft and an H of 2 ft. Weir Y has a B of 3 ft and a H value of 1 ft. Assume that the error in H is more or less constant and equally difficult whatever the head on the weir.

3-29 The potential in volts E of an experimental battery is to be found from

$$E = E_0 + 0.08 \ln \frac{C_{ox}}{C_{red}}$$

where E_0 is -0.2 and without error, C_{ox} is the molar concentration of the oxidizer with value 5 mol/liter and probable error 0.6 mol/liter and C_{red} is the concentration of the reductant with value 1 mol/liter and probable error 0.2 mol/liter. What will be the percentage probable error in E?

3-30 The following three equations are to be used to process the results of a heat-transfer test on a condenser:

$$NTU = \ln \frac{1}{1 - e} \tag{3-31}$$

$$NTU = \frac{UA}{wC_p} \tag{3-32}$$

$$e = \frac{T_{c,out} - T_{c,in}}{T_h - T_{c,in}} \tag{3-33}$$

(a) If T_h is equal to $212°F$ with no error, $T_{c,in}$ equal to $60°F$ with an error of $\pm 1°F$, and $T_{c,out}$ is equal to $180 \pm 1°F$, find the error in e from Eq. 3-33.

(b) Obtain the error formula for NTU with a known error in e from Eq. 3-31 and use the results of the above computation to estimate the error in NTU (which is dimensionless).

(c) The purpose of the test is to find U in Eq. 3-32. With $w = 500$ lb/h ± 5 lb/h, $C_p = 0.25$ Btu/lb $\cdot$ $°F$ with no error, $A = 50$ ft^2 with no error, what will be the error in U including the already found error in e? Comment on improvements in the experiment.

3-31 Refer to Prob. 2-31 involving a double-peaked distribution. What would be the propagated error if this distribution's standard deviation were combined with that of the distribution pictured in Fig. 2-15a in a sum-type (Eq. 3-27) function? Now suppose that the standard deviation of Prob. 2-31 were obtained from its probable error, assuming that the deviation was normal (using Eq. 3-28). What does s_r become, and what is its error? What is the equivalent of Eq. 3-29 for the distribution of Prob. 2-31?

3-32 The accompanying figure shows an apparatus consisting of a duct flowing air containing an unknown vapor concentration (expressed in milligrams per liter) with symbol C_{max} and an absorbent chemical that gradually extracts the vapor from the flowing air, following the equation $\Delta C = kC_{max}e^{-b/\Delta\theta}$, where ΔC is the weight of vapor in chemical after time $\Delta\theta$ has passed, k is a known constant (value 1.0) with units liter^{-1}, and b is a known system constant of value 3 s. The system will be used to determine C_{max} by measuring ΔC and $\Delta\theta$. The probable error in the time measurement is about 0.2 s, and for ΔC the error is 2 g. We wish the determinations to be made as fast as possible, and we anticipate that C_{max} will be in the neighborhood of 20 g/liter. What is a good operating plan? What probable error will occur in C_{max}? How can we improve the speed and accuracy of the determinations?

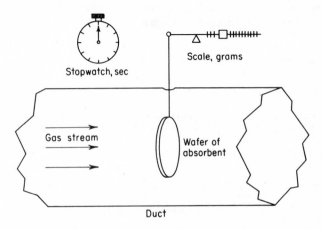

Stopwatch, sec

Scale, grams

Gas stream

Wafer of absorbent

Duct

3-33 A mixture of gases in unknown proportions is to be tested for its mixture gas constant R/M to be found from the perfect gas law $Pv = (R/M)T$, where P is the sample pressure (100 psia), T is the sample temperature (500°R), and v is the specific volume (1 ft^3/lb). We estimate the temperature error to have a 2-SD value of 10°R, and the pressure 2-SD error is 1 psia. The specific volume will be found by weighing a container with an exact internal volume of 1 ft^3. What must be the maximum error in weighing of this sample container if R/M is to be obtained with a 2-SD range of no more than 3%? If the container weighs 50 lb, what must be the percentage accuracy of the scale?

3-34 The downstream depth D_2 of a hydraulic jump is related to the upstream depth D_1 by

$$D_2 = \frac{D_1}{2}\left(-1 + \sqrt{1 + \frac{8Q^2}{gD_1^3b^2}}\right)$$

where Q is the volume flow rate, nominal 3 m^3/s, D_1 is the upstream depth, nominal 1 m; b is the channel width, nominal 1 m; and g is 9.8 m/s^2 with no error. If we wish to predict D_2, and Q, D_1, and b may all have error, find the sensitivity ratios of these three parameters and suggest where the most important error sources lie.

3-35 The virtual-mass experiment in Example 3-5 was repeated with the subject now standing upright and stepping off the board. C_dA is now 0.111 m² but with at least 10% uncertainty, F_b is the same, V_0^2 is now 35.86 m²/s², and the D_{max} values of seven trials are 2.31, 2.26, 2.11, 2.39, 2.36, 2.44, and 2.57, all in meters. Since virtual mass is now smaller (less water entrained) and the uncertainty of water entrance velocity is higher since the subject now enters progressively, we expect a greater variation and more uncertain results. What is M_v for this case, and how uncertain is it? Is it reasonable to state that M_v is *definitely greater* than 72.56 kg, the weight of the subject? Discuss this in terms of the SD of the error estimates as in Example 3-5 and also in terms of the SD of the mean as considered in Chap. 2.

BIBLIOGRAPHY

Berge, R. T.: *Am. Phys. Teacher,* vol. 7, pp. 351–357, 1939.

Deming, W. E.: *Statistical Adjustment of Data,* chap. 3, Wiley, New York, 1943.

Fleisher, H., and L. Olsen: *Am. J. Phys.,* vol. 18, pp. 51–52, 1950.

Hanau, R.: *Am. J. Phys.,* vol. 19, p. 382, 1951.

Hoehner, S. F.: *Fluid-Dynamic Drag,* published by the author, Midland Park, N.J., 1965.

Keenan, J. H., and F. G. Keyes: *Thermodynamic Properties of Steam,* Wiley, New York, 1936.

Kline, S. J.: Discussion of paper by Thrasher and Binder, *Trans. ASME,* vol. 79, no. 2, February, 1957.

Kline, S. J., and F. McClintock: Describing Uncertainties of Single Sample Experiments, *Mech. Eng.,* vol. 75, p. 3, January, 1953.

Rowles, W.: *Am. J. Phys.,* vol. 27, pp. 62–63, 1959.

Schenck, H.: *Case Studies in Experimental Engineering,* McGraw-Hill, New York, 1970.

Spiegel, M. H.: *Probability and Statistics,* Schaum's Outline Series, McGraw-Hill, New York, 1975.

Volk, William: *Applied Statistics for Engineers,* chap. 7, McGraw-Hill, New York, 1958.

Wilson, W. A.: Design of Power-Plant Tests to Insure Reliability of Results, *Trans. ASME,* vol. 77, no. 4, pp. 405–408, May, 1955.

Worthing, A. G., and J. Geffner: *Treatment of Experimental Data,* chap. 9, Wiley, New York, 1943.

FOUR

REDUCTION OF VARIABLES: DIMENSIONAL ANALYSIS

The study of individual uncertainty and uncertainty combinations, as discussed in the previous two chapters, is certainly the most crucial and imperative aspect of experimental planning. If potential errors are excessive, no amount of ingenious planning or advanced statistical "fixes" will accomplish very much. If, however, the uncertainty analysis shows that all is well, the investigator can then turn to more refined aspects of planning technique. The purpose of such planning can be simply stated: it is to obtain the maximum amount of useful data under the best possible control with a minimum expenditure of operating and calculating time.

As we shall see, there are several quite simple ways in which a given test can be made compact in operating plan without loss in generality or control. The best-known and most powerful (for the engineer) of these is *dimensional analysis*. Some 50 years ago, dimensional analysis was used primarily as an experimental tool and specifically as a means whereby several experimental variables could be combined to form one. The fields of fluid mechanics and heat transfer benefited greatly from the application of this tool, almost every major experiment in these areas being planned with its help. As the technique gradually became a part of engineering curricula, the original purpose behind the methods slipped away. Instead of a principle whereby modern experimenters can substantially improve their working techniques, dimensional analysis is frequently presented as a pedagogical device or a historical curiosity. It is often applied to experimental problems dating back 30 years or more, or else serves as a neat and rapid way of deriving certain functional relationships without recourse to complex theory.[*]

---

[*]For some excellent examples of this side of dimensional analysis see Conn and Crane, 1956.

As a result, young engineers often look on dimensional analysis as no more than an ingenious way of getting some already known results, or as a historical curiosity having little relevancy to modern technology. It is hoped that this chapter will in some measure counter such ideas by presenting this technique as it was originally conceived, namely, as a method whereby many experiments can be made shorter without loss of control.

4-1 THE BUCKINGHAM THEOREM

To apply this method properly, the investigator must know exactly the kind and number of *fundamental variables* in the test. A fundamental variable we shall define as any experimental variable that influences the test and can be changed or altered independently of the other test variables. Fundamental variables must be distinguished from *controlled variables*. We could, for example, change the acceleration of gravity independently of all other test variables by moving the apparatus to the moon (so it is certainly fundamental), but we realize that, practically, this is impossible for the usual laboratory test.

Assuming that the experimenter does know all the variables, their number can immediately be reduced through the application of the first part of the *Buckingham theorem*, which states: If any equation is dimensionally homogeneous, it can be reduced to a relationship among a complete set of dimensionless products.

A *dimensionally homogeneous* equation is one whose form does not depend on the fundamental units of measurement. An example is the familiar Fanning friction-factor equation,

$$\Delta P = f \frac{L}{D} \frac{V^2}{2g} \tag{4-1}$$

where the variables might be in units of feet and seconds, meters and hours, rods and minutes, or any consistent system. Conversely, the equation relating the heat flow per unit area g/A from a radiating body at temperature T as proposed by Dulong and Petit, $g/A = C(1.0077)^T$, is not homogeneous because using T in degrees Kelvin will require a very different function from using Rankine temperatures. The correct formula for this situation was found later to be $g/A = \sigma T^4$, with σ being a dimensional constant. Indeed, we should doubt that any natural occurrence can possibly be explained by a nonhomogeneous equation, except as a temporary or approximate expedient.

The *dimensionless products* noted in Buckingham's rule are simply products and quotients made up of the variables such that the dimensions cancel in each group. In the case of the Fanning equation, we can write it in terms of three dimensionless products or groups: $\Delta P/(V^2/2g)$, f, and L/D. The Buckingham theorem is by no means so trivial as might appear from this simple example, nor is its proof obvious.[*]

We have already suggested that nonhomogeneous equations cannot represent the complete mathematical statement of natural occurrence. We may not be able to recognize all the variables that influence a test, but we should realize that they and their dimensionless equation have reality regardless of whether it is apparent. Failure to obtain a set of dimensionless products is thus a sure sign that something is missing.

In the most general form of the Fanning equation, we note that ΔP is usually the quantity of interest. We see that it is a function of the pipe length L, the diameter D, and the flow velocity V, all of which are individually variable. The gravitational acceleration g is not easily changed, but it must be included. More thought should convince us that the fluid properties of density and viscosity are independently variable (by changing the type of fluid or its temperature), and an examination of various pipe interiors will show that roughness height e is also variable. This makes eight fundamental variables, and we can write the general equation

$$\Delta P = \phi(L, D, V, p, u, e, g) \qquad (4\text{-}2)$$

where ϕ means "a function of."

Buckingham's theorem states that this functional relationship (if homogeneous) can be written in terms of dimensionless products. We know from long experience that these can be

$$\frac{\Delta P}{V^2/2g} = \phi'\left(\frac{L}{D} \; \frac{VDp}{u} \; \frac{e}{D}\right) \qquad (4\text{-}3)$$

which can be seen to be dimensionless if consistent units are used throughout. For the experimentalist, finding the function ϕ' in Eq. 4-3 is far easier than finding the function ϕ in Eq. 4-2. Instead of varying each of seven variables in turn, some of which may be hard to manipulate, the investigator need only vary each of three groups. This is experimentally a great simplification and makes plotting and data analysis far quicker and more accurate.

---

[*]For an interesting geometric proof see Corrsin, 1951. Most proofs involve determinants; see Bridgman, 1931.

Let us see how the groups in Eq. 4-3 can be obtained in a simple and systematic way. We shall follow the so-called Rayleigh method of solving dimensional systems, and we begin by writing the variables in the friction loss system in terms of three *fundamental dimensions*, mass M, time θ, and length L. A list of the basic dimensional formulas will be found in Appendix B.

Name of variable	Symbol	Dimensional formula
Head loss in pipe	ΔP	L
Length of pipe	L	L
Pipe diameter	D	L
Fluid velocity	V	L/θ
Fluid viscosity	μ	$M\theta^{-1}L^{-1}$
Fluid density	p	ML^{-3}
Roughness height	e	L
Acceleration of gravity	g	$L\theta^{-2}$

We now assume that there exists a relationship between these quantities such that

$$(L^a, D^b, V^c, \mu^d, p^e, e^r, g^g) = \Delta P \qquad (4\text{-}4)$$

Now Eq. 4-4 can be rewritten with the dimensional formulas of the above table inserted,

$$\phi(L^a, L^b, (L\theta^{-1})^c, (M\theta^{-1}L^{-1})^d, (ML^{-3})^e, L^f, (L\theta^{-2})^g) = L$$

If this equation is to be dimensionally homogeneous, the following relationship among the exponents must apply:

For M: $0 = d + e$

For L: $1 = a + b + c - d - 3e + f + g$

For θ: $0 = -c - d - 2g$

We have three equations and seven unknowns. Let us simplify as far as possible, eliminating e, c, and b. Then $e = -d$, $c = -d - 2g$, and $b = 1 - a - d + g - f$. These equations can now be substituted for the exponents in Eq. 4-3,

$$(L^a, D^{1-a-d+g-f}, V^{-d-2g}, \mu^d, e^f, g^g) = \Delta P$$

Now, collecting terms with like exponents, the dimensionless groups are easily formed,

$$\phi\left(\left(\frac{L}{D}\right)^a, \left(\frac{VDp}{\mu}\right)^{-d}, \left(\frac{e}{D}\right)^f, \left(\frac{Dg}{V^2}\right)^g\right) = \frac{\Delta P}{D} \tag{4-5}$$

The original eight variables of the problem are reduced to five groups, and the investigator has gone as far as possible in this problem with dimensional analysis. Now we should begin to test for the actual function that includes these groups and expresses the behavior of the pipe friction system. Experiments in the laminar region will soon reveal the following function:

$$\frac{\Delta P}{D}\left(\frac{Dg}{V^2}\right)^1 = 32\left(\frac{L}{D}\right)^1 \left(\frac{VDp}{\mu}\right)^{-1} \left(\frac{e}{D}\right)^0$$

(Later chapters will consider some of the means by which this could be discovered from test data.) Thus, the final result is the familiar equation for laminar friction loss in a round pipe:

$$\frac{\Delta P}{V^2/2g} = \frac{64}{N_{re}}\frac{L}{D} \tag{4-6}$$

Only three groups are thus needed here (four in turbulent flow), but we could not infer this from dimensional reasoning. Still, the experimental simplification is obvious.

4-2 THE PI THEOREM

Let us consider another fluid system involving a submarine of characteristic dimension d moving at various speeds in viscous fluid and impeded by a drag force D, as shown in Fig. 4-1. Again, we shall use the same fundamental dimensions of mass, time, and length, so that the table of variables becomes:

Name of variable	Symbol	Dimensional formula
Fluid velocity	V	$L\theta^{-1}$
Characteristic dimension	d	L
Fluid density	ρ	ML^{-3}
Fluid viscosity	μ	$M\theta^{-1}L^{-1}$
Drag force	D	$ML\theta^{-2}$

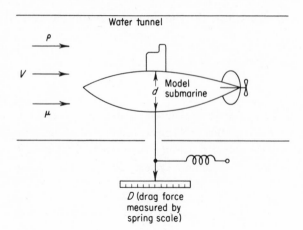

Figure 4-1 The diagram of a model submarine experiment showing the variables expected to influence this problem.

Since we have chosen a system of fundamental dimensions made up of mass, time, and length, we have derived the dimensions of force from these basic dimensions and the equation of Newton's second law. As we shall note later, force can be taken as a fundamental dimension if the analyst so chooses. As before,

$$\phi(V^a, d^b, \rho^c, \mu^d) = D$$

and

$$\phi((L\theta^{-1})^a, L^b, (ML^{-3})^c, (M\theta^{-1}L^{-1})^d) = ML\theta^{-2} \tag{4-7}$$

So the exponent equations become

For M: $\qquad\qquad c + d = 1$

For L: $\qquad\qquad 1 = a + b - 3c - d$

For θ: $\qquad\qquad -2 = -a - d$

Solving in terms of d only,

$$c = -d + 1 \qquad b = 2 - d \qquad \text{and} \qquad a = 2 - d$$

and Eq. 4-7 becomes

$$\phi(V^{2-d}, d^{2-d}, \rho^{1-d}, \mu^d) = D$$

and

$$\frac{D}{\rho V^2 d} = \phi'\left(\frac{\mu}{V\alpha\rho}\right)$$

or, in its more usual form,

$$C_d = \phi'(N_{re}) \tag{4-8}$$

where C_d is the familiar drag coefficient, which is a function of the Reynolds number only. Thus, to represent the behavior of this submarine, we need not plot drag versus V, d, ρ, and μ separately, but need only find a single curve of C_d versus N_{re}. It should be obvious that many fewer tests points are needed for this latter job, as Fig. 4-2 reveals.

Notice, however, that the resulting curve will hold only for submarines that

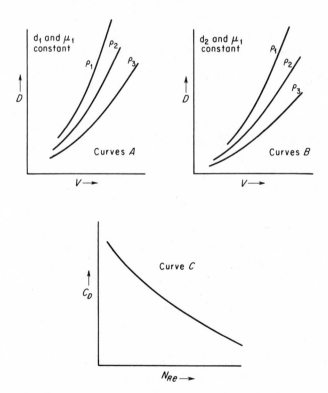

Figure 4-2 Without dimensional analysis, many curves like A and B would be required to express the submarine data. With dimensional analysis, only a single curve (C) is necessary.

are *geometrically similar.* Our single characteristic dimension d can only specify the overall size of the boat, and it can tell us nothing about its shape, taper, streamlining, and so on. To obtain a general expression for submarines of any shape would require a huge number of dimensional ratios and a test effort too vast to consider. Thus the investigator must beware of drawing too extensive conclusions from dimensional analysis or trying to apply the results too liberally.

Both the pipe friction and the drag problem illustrate the second part of Buckingham's theorem, which is useful in checking the results of a dimensional analysis. This is the *pi theorem*, which states: If there exists a unique relation $\phi(A_1, A_2, \ldots, A_n) = 0$ among n physical quantities that involve k primary dimensions, then there also exists a relation

$$\phi'(\pi_1', \pi_2', \ldots, \pi_{n-k}) = 0$$

among the $n - k$ dimensionless products made up of the As. Both the previous examples follow this rule. The friction case had eight physical quantities n, and we chose three primary dimensions k so that the pi theorem would predict $8 - 3$ or five dimensionless products of πs. This is mathematically what was obtained, although experimental work then showed that only three or four were actually required. In the drag problem there are five variables and three primaries yielding two πs as found, both of which, it happens, are necessary.

4-3 SELECTION OF GROUPS AND VARIABLES

At first glance, dimensional analysis and the Rayleigh analytic method of finding groups seem relatively automatic in application, but such, unhappily, is often not the case, particularly when one is dealing with new and strange experiments. Consider our submarine problem again. Suppose that we had solved these three equations in terms of c instead of d. We then obtain

$$d = -c + 1 \qquad a = c + 1 \qquad \text{and} \qquad b = c + 1$$

Putting these values in Eq. 4-7 as before, we obtain

$$\frac{D}{Vd\mu} = \phi'' \left(\frac{Vd\rho}{\mu} \right) \tag{4-9}$$

Equation 4-9 is surely as correct as Eq. 4-8. The groups are dimensionless, and either equation checks the pi theorem. To the experienced investigator, however,

Eq. 4-9 is not as useful as Eq. 4-8. The group $D/\rho V^2 d^2$ includes the drag force, the kinetic energy of the flowing fluid, and a dimension squared that suggests an area, perhaps a frontal area on which the fluid impinges. Thus, this group could be taken as the ratio of the actual drag to the force caused by impact of the fluid on the submarine frontal area. The group $D/Vd\mu$, on the other hand, is not an obviously meaningful collection of variables. The combination of viscosity with a length and velocity is unusual and not especially significant. Another group of little use can be obtained by solving the exponent equations for a, getting $\rho D/\mu^2$. If this group contained only fluid properties, it might be of interest, but it also includes the drag force and is thus of questionable utility.

The point, then, is that there are several solutions to many dimensional systems, and that, although all solutions may be correct, they are not equally useful.

The reader may have begun to wonder about the inclusion or exclusion of certain variables in an experimental system. Why, for example, did we include the acceleration of gravity in the friction problem and not in the drag-system analysis? Suppose that we miss some important variable? Suppose that irrelevant variables are mistakenly included?

Consider a system of two bodies in open space, revolving around each other owing to mutual gravitational attraction. We suspect that the following variables might specify the behavior of this system:

Name of quantity	Symbol	Dimensional formula
Mass of body 1	M_1	M
Mass of body 2	M_2	M
Distance of separation	R	L
Period of revolution	P	θ

Then we write

$$\phi(M^a, M^b, L^c) = \theta$$

but it is obvious that an error has occurred. Time appears on one side of the equation but not on the other, and a dimensionless number is not possible. This could mean that the revolution period is not an important variable, but since this is the variable we wish to find, such an explanation must be rejected. Either this is a case of a nonhomogeneous equation, or (more likely) we have left out something important. Any astronomer can quickly supply the missing item. It is the gravitational constant G, with units $M^{-1}L^3\theta^{-2}$. Including this in our analysis, we could easily obtain

$$\frac{PG^{1/2}M_2^{1/2}}{R^{3/2}} = \phi'\left(\frac{M_2}{M_1}\right) \tag{4-10}$$

and, with almost no knowledge of celestial mechanics, we have found that the revolution period of a double-star system is directly proportional to the separation distance to the $\frac{3}{2}$ power, a fact that is certainly not intuitively obvious. But, the reader may wonder: Why do we put a universal constant like G in the problem when it is by no stretch of the imagination a variable? No really satisfying answer is possible, but we might say that a universe having a G value less than this with nothing else changed would show a longer period of revolution. That no such universe probably exists is philosophically interesting but of no concern in dimensional analysis. *We require G because the system is not completely described without it.*

In the double-star system, the failure to include an important variable was immediately obvious. Let us now consider a circular disk of diameter d and negligible thickness, rotating in the center of a fluid-filled cylindrical casing completely enclosed and filled with a viscous liquid. Figure 4-3 shows the various important variables.

Name of quantity	Symbol	Dimensional formula
Disk diameter	d	L
Internal diameter	D	L
Fluid viscosity	μ	$ML^{-1}\theta^{-1}$
Clearance	c	L
Angular velocity	w	θ^{-1}
Torque	T	$ML^2\theta^{-2}$

We have six variables and three primaries; so we require three groups. The system is quite easily handled by *inspection*, without recourse to the formal Rayleigh method. For example, one group will logically be D/d, a second group might be D/c, and the third group could be $d^3w\mu/T$. When groups are found by inspection, it should be obvious that *all variables must appear at least once.* Then

$$\frac{T}{d^3w\mu} = \phi\ \frac{D}{d}, \frac{D}{c}$$

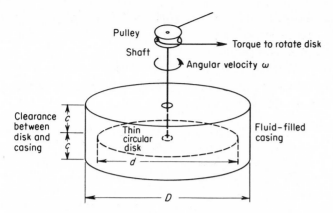

Figure 4-3 A diagram of a possible experiment involving a rotating thin disk in an enclosed viscous fluid, showing some of the nomenclature.

This might, at first glance, seem all right, but we should be suspicious. Motion in or out of a fluid almost always yields a Reynolds number, for which we need density. Looking at our system, we should be able to see that fluids with different densities will surely give different results even if all else is the same. Let us then add this new variable and rewrite the groups in a more logical fashion:

$$\frac{T}{\rho(wd)^2 D^2 c} = \phi\left(\frac{wD^2\rho}{\mu}, \frac{D}{c}, \frac{D}{d}\right) \tag{4-11}$$

The equation now seems to make more sense. A Reynolds number based on rotary velocity appears. The term $\rho(wd)^2$ is related to the fluid kinetic energy at the disk edge, whereas $D^2 c$ is easily transformed into the total volume of the container. Thus the denominator of the torque group can be "built" into some sort of kinetic energy term based on a reference lip velocity. We might wonder about the two groups D/c and D/d (which could just as well be c/d and c/D, and so on). Are both these groups needed to specify the system behavior completely? Although we might suspect that they are not, we cannot assume so on the basis of dimensional analysis. The pi theorem demands four groups, and experimentation must take over at this point.

The reader may now wonder: What about gravitational acceleration g? Surely, if the fluid did not fill the container so that a free surface remained, the position of the apparatus in the gravity field would be crucial. But if, as we assumed, the fluid fills the container, we suspect that gravity has little effect on the motion of the disk or fluid. Whether to insert g is a major puzzle, and real insight is needed if the proper decision is to be made. In the frictional system, we included g because the head loss ΔP was given in feet of the flowing fluid—feet, that is, in a gravity field of value g. In the drag problem, we sensed that the drag is independent of local gravity.

Consider now a string of length R connected to a fixed point with a rock of mass M at the other end. The rock whirls with a constant velocity, and we are interested in the force F in the spring (Fig. 4-4):

Name of quantity	Symbol	Dimensional formula
Mass of rock	M	M
Linear velocity of rock	V	$L\theta^{-1}$
String length	R	L
Acceleration of gravity	g	$L\theta^{-2}$
Force in string	F	$ML\theta^{-2}$

Then

$$F = \phi(M^a, g^b, V^c, R^d)$$

and

$$ML\theta^{-2} = \phi(M^a, (L\theta^{-2})^b, (L\theta^{-1})^c, L^d)$$

For M: $\qquad\qquad\qquad a = 1$

For L: $\qquad\qquad\qquad 1 = b + c + d$

For θ: $\qquad\qquad\qquad -2 = -2b - c$

and

$$a = 1 \qquad c = 2 - 2b \qquad \text{and} \qquad d = -1 + b$$

This gives a grouping of

$$\frac{FR}{MV^2} = \phi'\left(\frac{gR}{V^2}\right) \tag{4-12}$$

Let us imagine that we run an experiment with rocks of different masses whirling at different speeds and then plot our results. We would soon find (as any elementary physics student should suspect) that the plot in Fig. 4-5 results. The

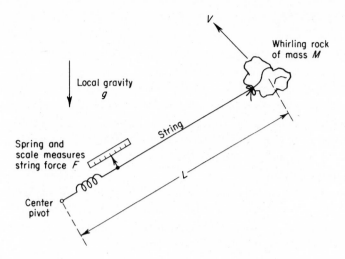

Figure 4-4 A whirling-rock system with some of the possible variables.

straight horizontal line shows simply that we have one too many variables. Our error here was the inclusion of g in our table of variables. A little more thought might convince us that the acceleration of gravity has little effect on this system—the force in the string will be the same on the moon as on the earth. Leaving out g, we obtain a single group, so that

$$\frac{FR}{MV^2} = \text{const} = k \tag{4-13}$$

is the complete relationship, and only k remains to be determined. (It would be a rather poor engineer who would need tests to find k in this case.)

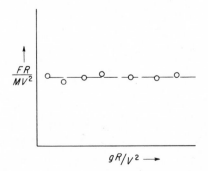

Figure 4-5 The probable result of an experiment on whirling rocks using the variables noted in Fig. 4-4.

4-4 A STEPWISE METHOD

The Rayleigh method of finding dimensionless groups is often extended by the application of determinants, particularly where the exponent equations are complex. Through determinants, a fuller understanding of the intricacies and pitfalls of dimensional analysis may also be gained. A newer and equally powerful method is described by Ipsen[*] and called by him the "step-by-step" approach. In this method the primary dimensions are "eliminated" through variable combinations in a sequential fashion. The method is quite simple and can best be followed through example.

Suppose that we are running tests on a series of identical pumps having characteristic impeller diameters D, impeller rotational speed N, fluid density ρ, and fluid volumetric flow rate Q. The dependent or measured test variable is to be ΔP, the pressure rise in the pumped fluid. Now we write

$$\Delta P = f(Q, N, \rho, D) \tag{4-14}$$

which in the $ML\theta$ system is dimensionally

$$\frac{M}{L\theta^2} = f\left(\frac{L^3}{\theta}, \frac{1}{\theta}, \frac{M}{L^3}, L\right)$$

Now, if Eq. 4-14 expresses the true experimental situation,

$$\frac{\Delta P}{\rho} = f(Q, N, \rho, D) \tag{4-15}$$

is equally correct and general. But we have eliminated the primary dimensions M from all the variables but ρ itself. Thus ρ cannot have a proper place in Eq. 4-15, because it is the only term now having the dimension M. Equation 4-15 should therefore be written

$$\frac{P}{\rho} = f(Q, N, D) \tag{4-16}$$

or dimensionally,

$$\frac{L^2}{\theta^2} = f\left(\frac{L^3}{\theta}, \frac{1}{\theta}, L\right)$$

[*]Ipsen, 1960, chap. 11.

We now can eliminate the time dimension θ in a similar fashion using the variable N:

$$\frac{\Delta P}{N^2} = f\left(\frac{Q}{N}, N, D\right) = f\left(\frac{Q}{N}, D\right) \tag{4-17}$$

which in dimensional form is

$$L^2 = f(L^3, L)$$

Finally, using D will eliminate the L dimension,

$$\frac{\Delta P}{N^2 \rho D^2} = f\left(\frac{Q}{ND^3}\right) \tag{4-18}$$

showing that a dimensionless "pressure-boost term" will be some experimental function of the important dimensionless pump parameter Q/ND^3.

In this case, we were able to eliminate the primary dimensions by using the original variables. This is not always possible and is, in any case, not necessary, as the first example in the next section will show.

The Ipsen stepwise method is not quite so easy as may appear here, especially when many variables and four or five primary dimensions are involved. It has the great advantage, however, of giving the analyst very good control over the form of the groups, both through selection of the sequence with which the primaries are eliminated and the choice of just which variable will be used to eliminate a particular dimension.

4-5 THE CHOICE OF PRIMARY DIMENSIONS

So far we have assumed that the dimensions, mass, length, and time, are sufficient to construct all the needed system variables. Suppose that we were to take the previous example of the pump test and construct its experimental variables in an $MLV\theta$ system, V now being the *primary dimension*, volume, and appearing wherever we had L^3 before. The dimensions of Eq. 4-14 now become

$$\frac{M}{L\theta^2} = f\left(\frac{V}{\theta}, \frac{1}{\theta}, \frac{M}{V}, L\right)$$

Using the stepwise method as before, we first eliminate M to obtain Eq. 4-16 having the dimensional form

$$\frac{V}{L\theta} = f\left(\frac{V}{\theta}, \frac{1}{\theta}, L\right)$$

Next, we eliminate θ, as before, obtaining Eq. 4-17 with dimensions

$$\frac{V}{L} = f(V, L)$$

Now let us eliminate the V dimension using Q/N and obtain

$$\frac{\Delta P}{\rho N^2} \frac{N}{Q} = \frac{\Delta P}{\rho N Q} = f\left(\frac{Q}{N}, D\right) = f(D) \qquad (4\text{-}19)$$

and the dimensions are

$$\frac{1}{L} = f(L)$$

Notice that we have here used the combined variable Q/N to clear the equation of the primary dimension V, rather than using a single variable.

Finally, we must eliminate the L dimension with the D variable, obtaining

$$\frac{PD}{\rho N Q} = \text{const} \qquad (4\text{-}20)$$

Comparing Eq. 4-20 with Eq. 4-18, we seem to have made a dazzling improvement in our test program. We now need only a single point to specify completely the form of Eq. 4-20. Clearly, we have too many primary dimensions.

Some writers suggest the rule, "The dimensions chosen should be independent of each other," as a preventive of the kind of oversimplification that occurred in the pump analysis.[*] Such a rule tends also to oversimplify the problem of selecting the appropriate set of dimensions. Ipsen, for example, shows how including the primary dimension, angle, leads to an important simplification in the dimensionless-number set governing an electrical circuit system.[†] Angles

[*]Deutsch, 1962.
[†]Ipsen, 1960, pp. 176–177.

are usually regarded as the dimensionless function of two lengths, and are thus dimensionless numbers themselves.

In free-convection heat transfer, the $ML\theta T$ system of primaries (with T now as temperature) will yield the following groups:

$$\frac{hl}{k} = \phi' \left(\frac{L^2 \, \Delta T \, \rho k}{\mu^3}, \frac{L\mu \beta g}{k}, \frac{\mu C_p}{k} \right) \tag{4-21}$$

which checks the pi theorem $8 - 4 = 4$. Now if we use a system of five primaries $LM\theta TH$, where H is the quantity of heat instead of $ML^2\theta^{-2}$, we get

$$\frac{hl}{k} = \phi' \left(\frac{L^3 \, \Delta T \, \rho \beta g}{\mu^2}, \frac{\mu C_p}{k} \right) \tag{4-22}$$

which is the experimentally proved grouping for this type of test. Note that Eq. 4-22 is actually a special case of the more general Eq. 4-21 when the middle groups of Eq. 4-21 multiply directly.

The fact that using five primaries instead of four appears to give a "good" result in a heat-transfer system, whereas the use of four rather than three primaries in the pump analysis gave a "bad" result, indicates how cautious the analyst must be when selecting primary dimensions to describe a new or unfamiliar system.

Let us consider another heat-transfer system, this time a turbulently flowing fluid in a pipe gaining heat from the pipe walls. We assume that this system is specified by the variables:

Name of quantity	Symbol	Dimensional formula
Film coefficient	h	$H\theta^{-1}L^{-2}T^{-1}$
Diameter	D	L
Fluid velocity	V	$L\theta^{-1}$
Fluid density	ρ	ML^{-3}
Fluid viscosity	μ	$ML^{-1}\theta^{-1}$
Fluid specific heat	C_p	$HM^{-1}T^{-1}$
Fluid thermal conductivity	k	$HT^{-1}L^{-1}\theta^{-1}$

where we have chosen the $ML\theta HT$ system of primary dimensions. It will be left as an exercise to show that one possible set of groups is

$$\frac{hD}{k} = \phi \left(\frac{\rho VD}{\mu'}, \frac{\mu C_p}{k} \right) \tag{4-23}$$

A check with the pi theorem shows five primaries, seven variables, but three groups! This demonstrates an important limitation of the pi theorem, namely, that the theorem specifies only the *minimum* number of groups that make up a set. This is the number found in the majority of solutions. But the theorem does not prohibit more than this number of groups, as this example shows.

The presence of more than the pi-theorem-predicted number of groups results, in the stepwise method, when two primary dimensions are eliminated in a single step. Ipsen explains that the pi theorem is still operative but that the analyst is not using the *true minimum number* of Buckingham primaries.[*] In the case of Eq. 4-23, the $ML\theta T$ system will yield the identical groups, indicating that H is an extraneous primary.

Van Driest[†] suggests a modification of the pi theorem to take care of such situations. His rule is: The number of dimensionless products in a complete set is equal to the total number of variables minus the maximum number of these variables that will not form a dimensionless product. This maximum number is not particularly easy to determine in many cases. Taking the heat-transfer problem described by Eq. 4-23, we note that D, V, ρ, and h cannot be combined in any way to form a dimensionless group. Readers may convince themselves by trials that the addition of C_p or k to these four will, with appropriate exponential values, make a dimensionless group. We could add μ to D, V, ρ, and h and still have the dimension H left over, but μ with V, D, and ρ only can form a group so that the Van Driest rule would be broken. Thus we find seven variables minus a maximum of four that cannot form a group, leaving three groups, as the rule predicts.

This rule is particularly handy in problems involving mechanics, where a choice of an $ML\theta$ or an $FL\theta$ system must be made with mass equal to $F\theta^2 L^{-1}$. Being often without the θ dimension, $FL\theta$ systems are liable to a pi-theorem prediction of one number of groups, whereas the same system when analyzed by an $ML\theta$ system would be expected to produce a different number. The Van Driest rule will tell the analyst which is correct. The use of an $MLF\theta$ system might offer advantages in some cases, but care must be taken that the resulting groups are not oversimplified.

A special type of fundamental dimension may be included in systems involving electrical, magnetic, and atomic phenomena. Many variables found in these fields, such as charge, capacity, potential, and so on, are formed of the fundamental dimensions, mass, length, and time, plus either the dielectric constant K or the permeability μ. Speaking generally, these two fundmental

[*]Ipsen, 1960, pp. 173–174.
[†]Van Driest, 1946.

primaries express the electrical or magnetic properties of the space in which we are working.* The situation is somewhat analogous to dynamics, where we noted that force or mass could be used. The dielectric constant K in terms of an $ML\theta\mu$ system of primaries is $\mu^{-1}\theta^2 L^{-2}$, whereas the permeability in an $ML\theta K$ system is $K^{-1}L^{-2}\theta^2$.

Let us think of the energy per unit of volume T in an electromagnetic field of electric strength E and magnetic strength H. We must also include the permeability and either the dielectric constant or the velocity of electromagnetic waves in the medium c. K, μ, and c do not all have to appear since they are related by $K = (\mu c^2)^{-1}$, so that only two of the three have to be included in most electrical problems. (We have chosen electrostatic primaries.)

Name of variable	Symbol	Dimensional formula
Energy density	T	$ML^{-1}\theta^{-2}$
Electric field strength	E	$K^{-1/2}M^{1/2}L^{-1/2}\theta^{-1}$
Magnetic field strength	H	$K^{-1/2}M^{1/2}L^{1/2}\theta^{-2}$
Permeability	μ	$K^{-1}\theta^2 L^{-2}$
Wave velocity	c	$L\theta^{-1}$

Then

$$T = \phi(E^a,\ H^b,\ \mu^c,\ c^d)$$

and either method leads to one possible grouping

$$\frac{T\mu c^2}{E^2} = \phi'\left(\frac{H\mu c}{E}\right) \tag{4-24}$$

and although the pi theorem predicts only one group, two actually occur.

There is much more to the application of dimensional analysis in electrical and atomic phenomena than we can consider here. In dealing with microscopic systems, temperature is often not a primary but appears in terms of mass, length, and time (molecular kinetic energy). Certain quite basic numbers such as the "fine structure constant" can be studied through dimensional reasoning, but such matters are clearly beyond the interest and training of the engineering

---

*The engineer not familiar with these important ideas should reread those portions of a basic physics book dealing with the properties and theory of magnetic and dielectric materials, e.g., Sears and Zemansky, 1955, chaps. 27 and 35.

experimenter. The advanced and interested reader is referred to the Bibliography at the end of this chapter.

4-6 DIMENSIONAL ANALYSIS
AND THE DIFFERENTIAL EQUATION

When one is facing a system such as Fig. 4-3 for the first time, dimensional analysis is only one of the tools available for finding and combining the relevant variables. Most investigators will also take a try at writing the equations that govern the behavior of the system, for in so doing they may obtain a list of the experimental parameters and some idea of how they go together.

Consider Fig. 4-6, in which a small, buoyant sphere is dropped from a measured height and the maximum distance it goes below the surface is read from a submerged scale. In addition to the parameters shown in Fig. 4-6, we need the sphere diameter d, the fluid density ρ, the fluid viscosity μ, and the virtual mass of the sphere M_v, which is equal to the actual mass plus added mass of the entrained water.

Name of quantity	Symbol	Dimensional formula
Virtual mass of sphere	M_v	M
Entrance velocity	V_0	LT^{-1}
Fluid density	ρ	ML^{-3}
Fluid viscosity	μ	$M\theta^{-1}L^{-1}$
Maximum depth	D_{max}	L
Acceleration of gravity	g	$L\theta^{-2}$

Why is g here, since we specify V_0 as a parameter and not H? Imagine the experiment on the moon, with V_0 and the other parameters the same. Clearly the sphere would go deeper since the buoyant or restoring force is less. Now we write, using the Ipsen method,

$$D_{max} = f(M_v, d, V_0, \mu, \rho, g) \tag{4-25}$$

which in the $ML\theta$ system is dimensionally

$$L = f\left(M, L, \frac{L}{\theta}, \frac{M}{L, \theta}, \frac{M}{L^3}, \frac{L}{\theta^2}\right)$$

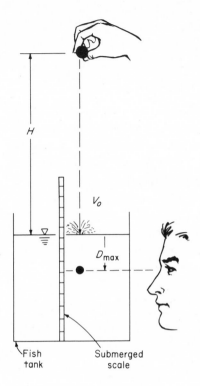

H

V_0

D_{max}

Fish
tank

Submerged
scale

Figure 4-6 Test apparatus for determining maximum submergence depth of a buoyant sphere.

We can eliminate the L dimension by combining the variable d with appropriate variables,

$$\frac{D_{max}}{d} = f\left(M_v, d, \frac{V\theta}{d}, \mu d, \rho d^3, \frac{g}{d}\right) = f\left(M_v, \frac{V_0}{d}, \mu d, \rho d^3, \frac{g}{d}\right)$$

or, dimensionally,

$$0 = f\left(M, \frac{1}{T}, M, \frac{1}{T^2}\right)$$

We eliminate M using ρd^3,

$$\frac{D_{max}}{d} = f\left(\frac{M_v}{\rho d^3}, \frac{V_0}{d}, \frac{\mu d}{\rho d^3}, \frac{g}{d}\right)$$

or, dimensionally,

$$0 = f\left(0, \frac{1}{\theta}, \frac{1}{\theta}, \frac{1}{\theta^2}\right)$$

and we can eliminate θ using V_0/d to get

$$\frac{D_{max}}{d} = f\left(\frac{M_v}{\rho d^3}, \; \frac{\mu}{V_0 d\rho}, \; \frac{gd}{V_0^2}\right) \tag{4-26}$$

This checks the pi theorem and has few surprises. We see a dimensionless depth parameter, a group that seems to relate the virtual mass to the mass of displaced fluid, the inverse of the Reynolds number based on entering velocity, and a Froude number. Now Eq. 4-26 is certainly a simplification of Eq. 4-25, but perhaps not much of one. To work out a series of tests with different spheres and different fluids to vary one of these numbers while holding the others constant would be difficult.

Let us write the differential equation of the sphere motion based on a free-body force balance on the sphere. After it strikes the water, we can identify three forces:

1. The deceleration force from Newton's second law is

$$-M_v \frac{d^2 D}{d\theta^2}$$

2. The drag force at any moment on the sphere is

$$-C_d \frac{\rho A}{2}\left(\frac{dD}{d\theta}\right)^2$$

3. And the buoyant force is the displaced fluid weight less the weight of the sphere or

$$-F_b$$

We call all three forces negative because they all resist the downward motion of the sphere. Since all forces on the body must be zero at any given instant,

$$M_v \frac{d^2 D}{d\theta^2} + \frac{AC_d\rho}{2}\left(\frac{dD}{d\theta}\right)^2 + F_b = 0 \tag{4-27}$$

Let us assume for a moment that this is too difficult to solve (which it is, if you do not change it a little). Suppose that we used Eq. 4-27 as our source of experimental parameters rather than the general or "guesswork" approach we have taken through most of the chapter. We already have one dimensionless variable C_d, as given in Eq. 4-8. Now, M_v is still there, but A, the sphere frontal area, replaces d, the diameter. The $dD/d\theta$ term shows that a velocity is needed; so we include V_0. The density ρ is included, and in place of g we have F_b, the buoyant upward force due to g acting on the displaced fluid. Viscosity is no longer needed, because C_d depends only on Reynolds number and thus viscosity. D_{max} must be included because it is the dependent variable.

Using the Ipsen method, it can be easily shown that this new list of variables leads to

$$\frac{D_{max}}{A^{1/2}} = f\left(\frac{M_v}{\rho A^{3/2}}, \frac{F_b}{\rho A V_0}, C_d\right) \qquad (4\text{-}28)$$

These dimensionless numbers look much odder than those in Eq. 4-26, but they are much closer to the correct function. If we substitute V for $dD/d\theta$ and $V(dV/d\theta)$ for $d^2D/d\theta^2$ in Eq. 4-27 and integrate, Eq. 3-24 is the result. Thus, the correct functional relationship among the dimensionless groups is

$$\frac{D_{max}A\rho}{M_v} = f\left(\frac{F_b}{\rho A V_0}, C_d\right) \qquad (4\text{-}29)$$

which we can obtain from Eq. 4-28 by combining the first two groups. This is not to say that obtaining Eq. 3-24 experimentally from Eq. 4-28 would not have its problems, but the difficulty would be nothing compared to trying to do it starting with Eq. 4-26. Figure 4-7 shows the correlation of data from three different spheres with the solid line showing Eq. 3-24. Deviations from theory result from C_d not being constant.

When facing a new experimental system, both mathematical and dimensional analysis can interact and play their parts.

4-7 SUMMARY

Although the methods of dimensional analysis certainly depend on sound physical insight into the particular problem, they are relatively automatic after the variables and primaries have been chosen. They can thus be easily summarized in a "cookbook" fashion, remembering always that this "recipe" requires the cook to choose the ingredients. All we can do here is give some mixing instructions.

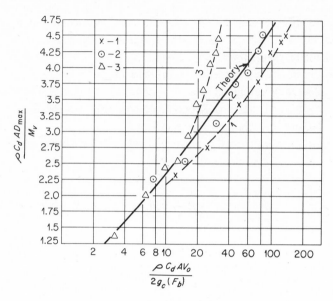

Figure 4-7 Dimensionless correlation of data from three spheres tested in the system of Fig. 4-6. The solid line shows the prediction of Eq. 3-24.

Step 1. Select those variables that are independent of each other and are believed to influence the system. Dimensional and natural constants must be included when they are judged to be significant. *This is the crucial step of the whole process.*

Step 2. Select a system of fundamental or primary dimensions in which all chosen variables can be expressed. Common systems are, for mechanics and fluid flow, $ML\theta$ or occasionally $FL\theta$; for thermal systems, $ML\theta T$ or $ML\theta TH$; for electrical and atomic systems $ML\theta K$ or $ML\theta \mu$ with temperature as an additional primary or expressed in terms of molecular kinetic energy.

Step 3. Write the dimensional formulas of the chosen independent variables and find the groups either by the formal methods or by inspection. The solution is correct if (a) every group is dimensionless, (b) at least as many groups appear as are predicted by the pi theorem, and (c) each variable appears at least once.

Step 4. Examine the resulting groups from the standpoints of common usage, physical significance, and (if least-squares analysis, Chap. 8, is to be used) restriction of test uncertainty to one group if possible. If the groups are not satisfactory from these criteria, we can (a) again solve the exponent equations until a better set of groups turns up, (b) choose another set of primary dimensions and start over, or (c) reexamine our choice of independent variables.

Step 5. When a satisfactory collection of groups is achieved, the investigator can plan the actual variation of the groups by varying selected variables in the apparatus. Actual test-run planning is discussed in Chap. 6.

PROBLEMS

4-1 The buckling load on a column is related to its length, diameter, and modulus of elasticity. How many primary dimensions are needed? What group(s) is (are) best for this system?

4-2 A box of volume V containing a liquid of density ρ is suspended on a spring of elastic constant k. We are interested in the period of oscillation of the box in the gravity field. Is g needed? Find the best groups using an $ML\theta$ system. Now repeat using an $MVL\theta$ system where V is used in place of L^3. Which system gives the best groups?

4-3 The Weber number is composed of length, velocity, density, and surface tension. What form does it take?

4-4 Gas bubbles rising in different fluids have their velocities measured and their diameters taken by photograph as they pass a given point D ft below the liquid-free surface. Decide which variables affect this system, and construct the proper groups.

4-5 In electricity, the Langmuir–Child formula relates the saturation current density i and the voltage in an evacuated diode V to the e/m value for the electron and the linear spacing between the electrodes x. Find the best group(s). How many test points would be required (minimum) to check this formula, and how are i and V in the tube related?

4-6 In the test of a centrifugal oil pump, a common practice is to plot efficiency and horsepower in or out versus flow rate at constant rpm. Using dimensional analysis and including the pump impeller diameter, obtain dimensionless groups involving (in turn, not all at once) power in, power out, and efficiency as a function of flow, revolutions per minute, fluid density, pressure boost, impeller size, and so on.

4-7 Rocket nose cones of characteristic shape D travel at various speeds in air of various pressures and temperatures. The tip temperature is measured. Construct groups. (*Note*: Do not omit the velocity of sound, which almost always appears in high-speed problems.)

4-8 A journal bearing is loaded with a force per unit length F. Important variables are diameter, bearing clearance, oil density and viscosity, oil pressure, and shaft rotation speed. Construct logical groups.

4-9 Ship propellers of given shape but different sizes are to be tested. Decided on logical groups after selecting the variables.

4-10 Satellite antennas of length L, diameter D, thermal conductivity k, surface emissivity e, and root temperature T lose heat to open space at absolute zero. The Stefan–Boltzmann radiation constant also must be included. Find the logical groups to fit a test program. (*Note*: The emissivity is a ratio of actual radiated energy to energy radiated by a black body at the same temperature.)

4-11 A pendulum swings in a viscous, imcompressible fluid. The variables are the swing period, the gravitational acceleration, the pendulum length, the swing angle, the bob diameter, the mass of the bob, and the fluid density and viscosity. The string is assumed to be very fine and weightless. Decide how many groups are needed and find a reasonable set by inspection of the variables.

4-12 In a structural test system, the following quantities are assumed to describe the experiment completely: beam stiffness (force per unit length), beam length, Young's modulus, a cross-sectional moment of inertia, and the unit density of the beam (in units of pounds mass per foot). Without performing any dimensional analysis or constructing any groups, decide whether this list is correct or whether there are too many or too few variables. (*Hint*: Write the variables in the $ML\theta$ and then the $FL\theta$ system. Which gives the most information?)

4-13 In the general case of a sphere moving in a compressible fluid, the following variables are assumed to govern the sphere's behavior: cross-sectional area of sphere A, fluid velocity V, fluid density ρ, fluid viscosity μ, fluid modulus of elasticity E, and drag of fluid on sphere D. Construct appropriate experimental groups and compare with the drag analysis in Sec. 4-2. What group drops out when the fluid is incompressible?

4-14 The *Cauchy number* is a possible group involved in the system described in Prob. 4-13. This group includes the modulus of elasticity E, the velocity V, and the fluid density ρ. What form does it take?

4-15 An *LCR* series circuit is described by the following variables: resistance R, capacitance C, inductance L, charge q, frequency f, and voltage V. Construct appropriate groups.

4-16 Surface waves on a liquid are influenced by the following variables: wave velocity V, gravitational acceleration g, wavelength λ, liquid depth D, liquid density ρ, and liquid surface tension σ. What would appropriate experimental groups be? An oceanographic engineer deals often with two special cases: (*a*) where surface tension is unimportant and the wave velocity is almost independent of wavelength ("shallow" wave) and (*b*) where surface tension is unimportant and the wave velocity is independent of depth ("deep" wave). What do the groups become in these two cases?

4-17 A long transmission line with a voltage suddenly applied to one end is governed by current I, voltage E, resistance per unit length r, inductance per unit length l, capacitance per unit length c, and time t. What test groups might apply to this system?

4-18 In an oceanographic study of beach movement, waves with deep-water period T_0 (seconds), length L_0, and height H_0 were found to move Q ft^3 of sand per second per foot of wave crest. The waves were also measured to find their energy in foot pounds per foot of wave crest, this energy being taken as relatively independent of the wave dimensions. What other variables are needed to construct groups to correlate the data?

4-19 A viscosimeter measures the amount of time t for a given amount of fluid of density ρ, viscosity μ, and tube length L to drain out. What other variable is needed? Show that for a given instrument, the time to drain is a function of the kinematic viscosity only.

4-20 The *Peclet number* is made up of fluid density, fluid specific heat, fluid thermal conductivity, the fluid velocity, and some characteristic dimension. What form does it take? The *Graetz number* replaces the density and velocity by a mass flow rate (pounds per second). What form does it take? How are the Peclet and Graetz numbers related for flow in a round tube of diameter D?

4-21 A triangular weir has a volume rate of flow Q, a width W, a height on the weir H, and we expect that water density, viscosity, and local gravity g will also effect the experiment.

(*a*) Find the groups.
(*b*) If we replace the variables W and H by θ and H, where θ is the angle of the weir, find the new groups.
(*c*) If viscosity does not play a large part, what is the result? How are Q and H related?

4-22 The depression of a liquid in a capillary tube y is a function of the surface tension of the liquid–vapor film s_{lv}, the density of the fluid ρ, the radius of the tube R, the angle between the tube wall and the film at the wall θ, and g. Decide how one might run the experiment to relate these variables.

4-23 The terminal velocity V of small spheres of diameter d, in an oil of density ρ and viscosity μ, under the influence of g, are to be studied. What other variable is needed? What groups result?

4-24 A hovercraft test outfit responds to variations in its base area A, its base pressure P, the air flow in volume per minute from the base Q, the mass of the system M, gravitational acceleration g, and the height off the floor H. We would prefer to vary only flow Q and base pressure. Suggest dimensionless groups for the test.

4-25 The normal stress S due to a shrink of a collar of inner diameter D_i, outer diameter D_o, and interference when cold I_c is also dependent on Young's modulus E. How many groups are needed to meet the requirements of dimensional analysis and what are they?

4-26 In a study of splash energy loss in fluids, the following variables are assumed to govern the system: surface tension S, density ρ, mass of splashed water per unit area M, energy per area absorbed by water E_a, and g. Form a set of groups. Suppose that we omit the surface tension variable on the basis that it is probably a small force. What is the result? Criticize the result on the basis of simple physical reasonableness.

BIBLIOGRAPHY

Bridgman, P. W.: *Dimensional Analysis,* Yale University Press, New Haven, Conn., 1931.

Conn, G. T., and E. Crane: Some Applications of the Methods of Dimensions to Atomic Physics, *Am. J. Phys.,* vol. 24, pp. 543–549, 1956.

Corrsin, S.: Simple Geometrical Proof of Buckingham's Pi Theorem, *Am. J. Phys.,* vol. 19, pp. 180–181, 1951.

Deutsch, R. A.: Dimensional Analysis, *Electro-Technol.,* vol. 70, no. 2, pp. 107–114, August, 1962.

Huntley, H.: *Dimensional Analysis,* Rinehart, New York, 1951.

Ipsen, D. C.: *Units, Dimensions, and Dimensionless Numbers,* McGraw-Hill, New York, 1960.

Langhaar, H. L.: *Dimensional Analysis and Theory of Models,* Wiley, New York, 1951.

Murphey, G.: *Similitude in Engineering,* Ronald Press, New York, 1950.

Schenck, H.: *Case Studies in Experimental Engineering,* McGraw-Hill, New York, 1970.

Sears, F. W., and M. W. Zemansky, *University Physics,* complete ed., Addison-Wesley, Reading, Mass., 1955.

Van Driest, E. R.: On Dimensional Analysis and the Presentation of Data in Fluid Flow Problems, *J. Appl. Mech.,* vol. 13, no. 1, pp. A-34–A-40, March, 1946.

Whyte, L. L.: Dimensional Theory, Dimensionless Secondary Quantities, *Am. J. Phys.,* vol. 21, pp. 323–325, 1953.

FIVE

INSTRUMENT LOADING
AND RESPONSE

Roger J. Hawks

In the previous chapters we have studied measurement systems from a very general viewpoint. We assumed that the system would have an error, either a random precision error or a bias error or possibly a combination of the two. These errors can be analyzed and the usefulness of the test readings determined. In addition, the readings can be combined in certain ways to make the experiment shorter or more accurate. Certainly these things should all be considered when selecting an instrument system for a test.

There is, however, more to selecting an instrument for an experiment than just a consideration of its accuracy. *An instrument used to measure a quantity should not itself affect the quantity being measured.* Thus, it is necessary to consider the *loading* that the instrument exerts on the measurement. This loading is determined by the *impedance* of the instrument.

The reading errors we have discussed so far are all errors that will not go away with the passage of time. Thus, these errors are *static characteristics* of the instrument; they tell us nothing about how the instrument responds to a change in the quantity being measured. Since almost all engineering experiments involve making changes in one or more of the variables, the measured quantity is always changing. The way in which an instrument responds to these changes is called the *dynamic response.* Serious errors in readings can result when using even the most accurate instrument if the instrument cannot respond to changes in the measured variable.

In this chapter we shall consider these two aspects of instrument system design: impedance and loading, and dynamic response. Both must be carefully considered when selecting an instrument for a test. Also, in order to ensure good results, the dynamic response must always be accounted for in the test plan and when conducting the experiment.

5-1 IMPEDANCE AND LOADING

All measuring systems involve an *interface* between the instrument and the medium being measured. For instance, the glass bulb of a thermometer is an interface between the instrument (the thermometer) and the medium being measured (the fluid whose temperature is desired). In order for the measurement to be made, some energy must be transferred across the interface. In this case, heat is transferred from the fluid to the thermometer. However, if too much energy is transferred across the interface, the properties of the medium will be changed and the quantity that we wish to measure will be disturbed by the process of making the measurement. This change in the quantity because of the measurement is called *loading error*. Even though the loading error can never be eliminated, it can be reduced. The loading effect of an instrument is usually described by the term *input impedance.*

Impedance is primarily an electrical term. It refers to the apparent electrical resistance of a circuit when measured between open terminals. The input impedance of a voltmeter is then the apparent resistance between the input terminals when they are left open or unconnected. Likewise, the output impedance of a piezoelectric transducer is the apparent resistance between the open output terminals of the device.

Since the objective is to minimize the power drain on the medium, the quantity being measured will determine whether high or low impedance is desirable. If we are making voltage measurements, the power will be given by

$$P = \frac{E^2}{R_m} \tag{5-1}$$

Thus, to minimize the power drain, we wish to have the input impedance as high as possible. Similarly, if we are measuring currents, the power drain is

$$P = I^2 R_m \tag{5-2}$$

and we minimize the loading by using a low-impedance instrument.[*]

---

[*]Sometimes the term *admittance* is used in this case. Admittance is the reciprocal of impedance and current measurements are made with high-admittance instruments.

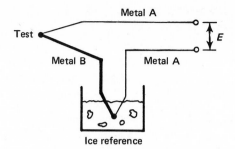

Figure 5-1 A thermocouple circuit used to measure temperature at the test junction.

Actually, what is important is not the input impedance itself, but rather how it relates to the output impedance of the circuit being measured. What we really need to do when measuring voltage is to make certain that the input impedance of the instrument is high when compared to the output impedance of the circuit.

To illustrate this, let us consider the measurement of the voltage produced by a thermocouple (Fig. 5-1). A thermocouple typically has an output voltage E on the order of micro- or millivolts. The output impedance of the thermocouple is just the resistance R_w of the lead wires. If we measure the output with a voltmeter, the circuit will look like Fig. 5-2, where R_m is the input impedance of the meter (the resistance of the coil in the meter movement). The voltmeter will read the voltage drop E_m across the meter. Now, since the current is the same throughout the circuit,

$$\frac{E}{R_w + R_m} = \frac{E_m}{R_m}$$

(5-3)

and the measured voltage will be

$$E_m = \frac{R_m}{R_m + R_w} E$$

(5-4)

In order for E_m to approximate E, R_m must be very large compared to R_w. Hence, the input impedance of the instrument must be high compared to the output impedance of the transducer.

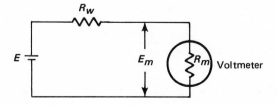

Figure 5-2 Output thermocouple circuit used to measure the voltage of the couple.

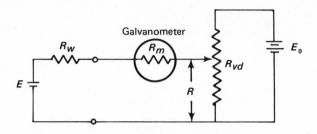

Figure 5-3 A potentiometer circuit that can be adjusted to have infinite input impedance.

In the case of the thermocouple, E is so very small that any error in E_m will be serious. Thus, thermocouple measurements should be made using a device that has infinite input impedance. Modern solid-state digital voltmeters can approach this ideal. *Potentiometer* circuits achieve the same result in a different manner.

In a potentiometer circuit (Fig. 5-3) the voltage-dividing resistor R_{vd} can be adjusted so that the voltage drop across R is exactly equal to E. Then there will be no current flowing through the meter, neither R_w nor R_m will affect the results, and the instrument will have infinite input impedance.

Example 5-1 The output voltage E of the circuit shown in Fig. 5-4 is to be read using the 200-V scale on a voltmeter with a 10,000 Ω/V rating. What is the loading error in this reading?

Solution Since moving-coil voltmeters use a galvanometer movement, the input impedance is proportional to the full-scale deflection. In this case,

$$R_m = 200 \times 10,000 = 2 \times 10^6 \ \Omega$$

The voltage measured by the voltmeter will be

$$E_m = \frac{R_m}{R_m + R_i} E$$

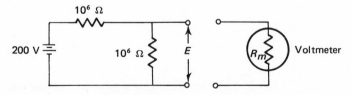

Figure 5-4 Example 5-1 circuit illustrating loading.

where R_i is the output impedance of the circuit. For this circuit,

$$\frac{1}{R_i} = \frac{1}{10^6} + \frac{1}{10^6}$$

or

$$R_i = 0.5 \times 10^6 \ \Omega$$

Thus

$$E_m = \frac{2 \times 10^6}{2.5 \times 10^6} E$$

$$= 0.8E$$

The meter will have a 20% error and will read 160 V instead of the correct 200 V. Notice that a 20% error resulted even though the input impedance of the meter was four times the output impedance of the circuit.

Although impedance is an electrical term, the concept can be generalized and applied to mechanical instruments as well. Mechanical impedance is sometimes called *stiffness* and is a measure of the resistance of the device to an applied force.* As in the case of electrical devices, the mechanical impedance of an instrument will need to be either large or small to minimize energy transfer depending on the nature of the quantity being measured.

An interesting example of mechanical impedance matching is the *soil pressure gauge* (Fig. 5-5) used to measure the subsurface pressures produced by applied loads. Such measurements are necessary for the proper design of highways, dams, and so on. In order to prevent undesirable loading of the soil by the gauge, the mechanical impedance of the gauge must be properly matched to

*The mechanical equivalent of admittance is called *compliance*.

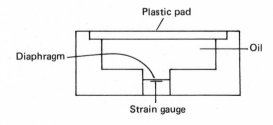

Figure 5-5 Diagram of a soil pressure gauge used to measure underground pressures.

the soil so that the gauge does not itself do work on the soil. Since the work done by a pressure is proportional to the change in volume, the pressure gauge should not cause any change in volume of the soil. If the gauge is too stiff, it will not deform as much as the soil, causing a reduction in soil volume and a local erroneous increase in pressure. Likewise, a gauge that is too compliant will easily deform, relieving the soil pressure. In this case, we must make the "input" impedance of the instrument exactly equal to the "output" impedance of the soil in order to minimize loading errors.

5-2 DYNAMIC RESPONSE

The dynamic response of an instrument is a description of how the instrument responds to changes in the inputs. Since we are now concerned with changes in the instrument output, we shall have to consider the differential equations that describe the instrument response. In doing this, we shall assume that the response of the instrument can be described by a *linear ordinary differential equation* with *constant coefficients.* As we shall see, this assumption is quite adequate for most instruments, but we must realize that nonlinear effects will often be present.

Instrument dynamic response can then be described by an equation of the type:

$$a_0 \frac{d^n y}{dt^n} + a_1 \frac{d^{n-1} y}{dt^{n-1}} + \cdots + a_{n-1} \frac{dy}{dt} + a_n y = bx(t) \qquad (5\text{-}5)$$

where y is the output of the instrument and x is the input. The output y might be an electrical signal such as a voltage, resistance, or current; or it could be a mechanical indication such as the displacement of a pointer, a scale, or a liquid surface.

Of course, the instrument response will depend on the nature of the input. An instrument might respond better to one type of input than to another. In order to compare instrument responses, we shall use three basic inputs.

Step Change in Input

A *step change* in the input occurs when the input variable instantaneously changes from one constant value to another (Fig. 5-6a). This takes place when a weight is placed on a scale or when a thermometer is placed in a container of liquid. An instantaneous change cannot actually occur in any physical quantity, but the changes are often rapid enough that the mathematical description of an

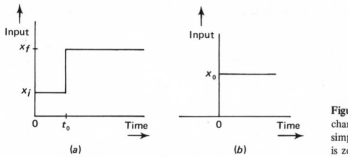

Figure 5-6 A step change in input (*a*) simplified such that x_i is zero at zero time (*b*).

instantaneous change will be sufficiently accurate. For purposes of analysis and comparison, we can take both x_i and t_0 to be zero as in Fig. 5-6*b*. This causes no change in the resulting concepts but simplifies the mathematics. The step input is described mathematically as

$$x = 0 \qquad t < 0$$

$$x = x_0 \qquad t > 0 \tag{5-6}$$

The input is not actually defined at $t = 0$. The instrument response to this input is called the step response.

Input Changing at a Constant Rate

In many instances, the change from one input level to another does not happen instantaneously. Instead the transition from one level to another occurs with a constant rate of change as in Fig. 5-7*a*. Such an input is called a *ramp input*.

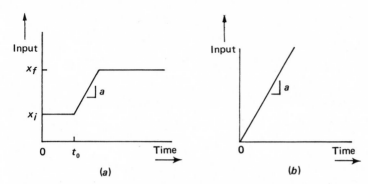

Figure 5-7 A ramp change in input (*a*) simplified such that the ramp begins at zero input and zero time (*b*).

Since the input is constantly changing, the instrument must dynamically respond in order to follow the circuit.

The most important parameter of this input is the rate of change or slope a. Thus, for comparison purposes, we simplify the input to that shown in Fig. 5-7b. This input is described mathematically by

$$
\begin{aligned}
x &= 0 & t &< 0 \\
x &= at & t &\geqslant 0
\end{aligned}
\tag{5-7}
$$

The response of an instrument to this input is called the ramp response.

Input Changing Sinusoidally

A *sinusoidally varying input* (Fig. 5-8a) rarely occurs in nature but is commonly found when making measurements on machinery. Oscillating inputs of this type are often used when conducting experiments. Oscillatory inputs are characterized by the period or frequency and by the amplitude of the oscillation. The *period* τ is the time between successive peaks of the input. The *frequency* is the reciprocal of the period, expressed in radians per second, that is,

$$
\omega = \frac{2\pi}{\tau}
\tag{5-8}
$$

The frequency of the input is usually called the *forcing frequency*. The amplitude is the maximum variation of the input away from the average. Hence:

$$
X = \frac{x_l - x_s}{2}
\tag{5-9}
$$

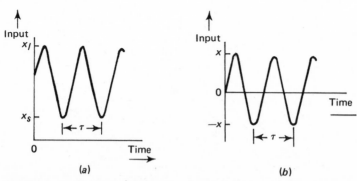

Figure 5-8 A sinusoidally varying input (a) simplified so that its excursions extend equally on each side of zero (b).

We usually assume that the oscillation occurs about a zero average as in Fig. 5-8*b*.

The sinusoidal input is described by

$$x = X \sin \omega t \tag{5-10}$$

The response of an instrument to this input is called the frequency response of the instrument.

5-3 RESPONSE OF ZERO–ORDER INSTRUMENTS

For purposes of determining the dynamic response, instruments are classified by the order of the differential equation used to describe the response. This is, of course, the highest derivative in the equation or the value of *n* used in Eq. 5-5.

A zero-order instrument is one whose response can be described by an algebraic equation. All zero-order instruments are described by the equation

$$y = Kx \tag{5-11}$$

All instruments behave as zero-order instruments for static inputs (*x* constant). Some also respond in this manner to dynamic inputs (*x* changing).

An example of a zero-order instrument is a wire strain gauge. The output of a strain gauge (the resistance change) is related to the input (the strain, ϵ) by the equation

$$\Delta R = GF\, R\epsilon \tag{5-12}$$

where *GF* is the *gauge factor* and *R* is the unstrained resistance of the gauge. Therefore

$$K = GF R \tag{5-13}$$

The constant *K* is called the *static gain* of the instrument. *K* determines the calibration of the device and, for a linear instrument, defines the sensitivity. Since the output of a zero-order instrument is always proportional to the input, there is *never* any error in the output due to the dynamic response.

Other examples of zero-order devices are the displacement potentiometer and the linear variable differential transformer (*LVDT*).

5-4 FIRST–ORDER INSTRUMENTS

A first-order instrument is described by the equation

$$a_0 \frac{dy}{dt} + a_1 y = bx \qquad (5\text{-}14)$$

This equation is usually divided through by a_1 and written:

$$\tau \frac{dy}{dt} + y = Kx \qquad (5\text{-}15)$$

K is again called the static gain and τ is the time constant since it has dimensions of time.

Most meteorological instruments are first-order devices. For example, consider the *cup anemometer,* Fig. 5-9. The wind is caught in the cups and exerts a torque on the shaft. This torque is proportional to the wind speed

$$T = bV \qquad (5\text{-}16)$$

The torque causes the anemometer to rotate. Friction in the bearings produces a retarding torque of cw. Thus, the equation of motion for the anemometer is

$$I \frac{d\omega}{dt} = bV - c\omega \qquad (5\text{-}17)$$

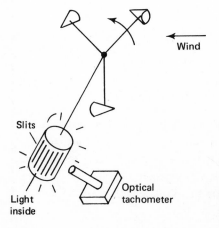

Slits

Light
inside

Optical
tachometer

Figure 5-9 Diagram of a wind-measuring cup anemometer using an optical tachometer to count revolutions.

or

$$\frac{I}{c}\frac{d\omega}{dt} + \omega = \frac{b}{c}V \qquad (5\text{-}18)$$

The angular velocity ω is measured by a light-chopping tachometer.

We see that the anemometer is a first-order instrument with a time constant of I/c and a static gain of b/c. It can also be shown that all temperature-measuring devices are first-order instruments.

Now let us look at the response of a first-order instrument to our basic inputs.

Step Response

The step response of a first-order instrument is given by the equation

$$\tau\frac{dy}{dt} + y = Kx_0 \qquad (5\text{-}19)$$

with the initial condition that $y = 0$ at $t = 0$. Equation 5-19 can be solved to give

$$y = Ce^{-t/\tau} + Kx_0 \qquad (5\text{-}20)$$

Using the initial condition gives

$$C = -Kx_0 \qquad (5\text{-}21)$$

so that the step response of the instrument is

$$y = Kx_0(1 - e^{-t/\tau}) \qquad (5\text{-}22)$$

This is plotted in Fig. 5-10 for several values of the time constant. We see that after a long time the output becomes equal to Kx_0. This explains why K was called the static gain. It plays the same role here that it did for a zero-order instrument. In addition, we see that the response is indeed that of a zero-order instrument if the input has not changed in value for some time, that is, has become static. Figure 5-10 clearly shows the effect of the time constant on the speed with which the instrument responds to a change in the input. Lower time constants result in faster responses. The initial slope of the response curve is the

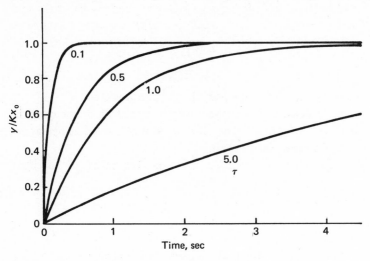

Figure 5-10 A plot of Eq. 5-22 for several values of time constant.

reciprocal of the time constant. We can determine the effect of the time constant on the output more easily by plotting y/Kx_0 as a function of t/τ as in Fig. 5-11. This universal graph applies to all first-order instruments. From Fig. 5-11, we see that the time constant is the time required to come to within $1/e$ times the static value, or 63.2%, of the true value. In order to be within 5% of the true value, we must wait three times the time constant before we read the instrument, and a

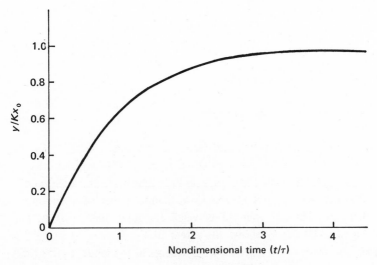

Figure 5-11 A replotting of Fig. 5-10 using a dimensionless time.

delay of 4τ is required to get only 2% error. This time delay because of the time constant is called the *response time* of the instrument. The response time is a major factor in any measurement taken with a first-order instrument. The response time should always be considered when selecting an instrument, and the test plans and procedures should be arranged to account for the response of the instrument.

Ramp Response

The ramp input is given by Eq. 5-7 so the response equation for a ramp input is

$$\tau \frac{dy}{dt} + y = Kat \tag{5-23}$$

The solution to this equation that satisfies the initial conditions is

$$y = Ka[t - \tau(1 - e^{-t/\tau})] \tag{5-24}$$

This is plotted in Fig. 5-12.

Since the input to the instrument is at, the true output should be Kat. Thus, the instrument response will be in error by the amount

$$e = y - Kat = Ka\tau(1 - e^{-t/\tau}) \tag{5-25}$$

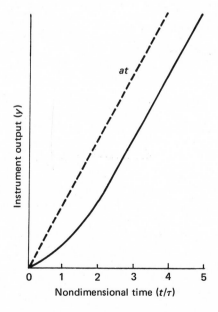

Figure 5-12 A nondimensional plot of Eq. 5-24 showing the response of a first-order instrument to a ramp input.

The second term of this error dies out with time. This is called the *transient error* and will be less than 0.05 in three time constants. However, the first term in the error $Ka\tau$ persists forever. This is the *steady-state error*. Since this error is proportional to the rate of change of the input a, it is also called *velocity error*.

The effect of the velocity error is to delay the output by the time constant. Thus, the first-order instrument does not read the current value of the input but rather the value that the input had τ seconds ago. It is therefore desirable to have a small value for τ.

Frequency Response

The frequency response of a first-order instrument is given by the equation

$$\tau \frac{dy}{dx} + y = KX \sin \omega t \tag{5-26}$$

Since the nonhomogeneous part of the equation is a sine function, the particular solution should be of the form

$$y_p = A \sin (\omega t + \phi) \tag{5-27}$$

Substituting into Eq. 5-26 results in

$$A \cos \phi - a\tau\omega \sin \phi = KX$$
$$A\tau\omega \cos \phi + A \sin \phi = 0 \tag{5-28}$$

From the second equation we get

$$\tan \phi = -\omega\tau \tag{5-29}$$

and solving simultaneously yields

$$A = KX \cos \phi \tag{5-30}$$

Since

$$\cos \phi = \frac{1}{\sqrt{1 + \tan^2 \phi}} \tag{5-31}$$

we find

$$A = \frac{KX}{\sqrt{1 + \omega^2 \tau^2}} \tag{5-32}$$

The frequency response of the first-order instrument therefore is

$$y = Ce^{-t/\tau} + \frac{KX}{\sqrt{1 + \omega^2 \tau^2}} \sin(\omega t + \phi) \tag{5-33}$$

where

$$\phi = \tan^{-1}(-\omega t) \tag{5-34}$$

The first term again represents a transient error, which dies out after a few time constants. The remaining steady-state error is what is usually referred to as the *frequency response*. A typical plot of the steady-state response will look like Fig. 5-13. The amplitude of the output is less than the static value KX. This is called *amplitude error*. The amplitude error gets worse as the frequency of the input is increased. The peak values of the output are also shifted in time from the input peaks. This is a *phase error*. The phase error also increases with frequency so that the output lags the input by a quarter cycle ($90°$) as ω approaches infinity.

The amplitude and phase errors can be plotted as in Figs. 5-14 and 5-15. Notice that both errors can be greatly reduced by reducing the time constant. Thus, a small time constant is again required in order to make accurate dynamic measurements.

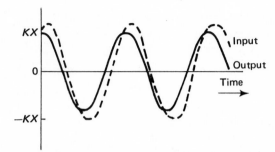

Figure 5-13 A typical example of instrument frequency response showing both amplitude and phase error.

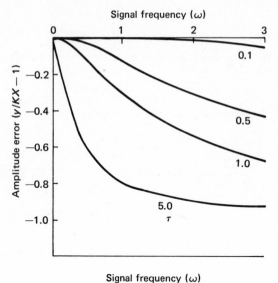

Figure 5-14 Typical amplitude errors shown as a function of signal frequency for several time constants.

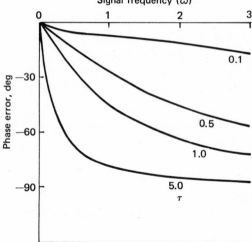

Figure 5-15 Typical phase errors shown as a function of signal frequency for several time constants.

Example 5-2 A resistance thermometer has a resistance of 25 Ω at $0°C$ and a sensitivity of 0.03 $\Omega/°C$. The time constant of the thermometer is 10 s. The thermometer is at $20°C$ when it is plunged into a tank of oil with a temperature of $170°C$. What is the percent error in the reading after 5, 15, and 30 s?

Solution Initially the resistance of the thermometer is

$$R_0 = 25 + 0.03(20) = 25.6 \ \Omega$$

The step response is

$$R - R_0 = K(T_f - T_0)(1 - e^{-t/\tau})$$

or

$$\frac{R - R_0}{T_f - T_0} = K(1 - e^{-t/\tau})$$

The correct final resistance is

$$R_f = 25 + 0.3(170) = 30.1 \ \Omega$$

Thus

t	$e^{-t/\tau}$	$(R - R_0)/(T_f - T_0)$	R	Error (%)
5	0.606	0.0118	26.77	−11.10
15	0.223	0.0233	29.09	−3.35
30	0.050	0.0285	29.87	−0.76

5-5 SECOND-ORDER INSTRUMENTS

A great many instruments respond according to the equation

$$a_0 \frac{d^2 y}{dt^2} + a_1 \frac{dy}{dt} + a_2 y = bx \tag{5-35}$$

These are called second-order instruments. The response equation is usually divided through by a_0 and written

$$\frac{d^2 y}{dt^2} + 2\rho \omega_n \frac{dy}{dt} + \omega_n^2 y = K\omega_n^2 x \tag{5-36}$$

K is the *static gain,* ρ is the *damping factor,* and ω_n is the *natural frequency.*

One example of a second-order instrument is the *d'Arsonval galvanometer*

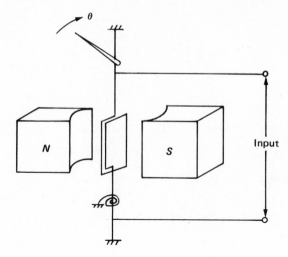

Figure 5-16 Diagram of a d'Arsonval galvanometer as an example of a second-order instrument.

(Fig. 5-16). If a current i flows through the coil, a torque is exerted on the coil equal to

$$T = bHi \qquad (5\text{-}37)$$

where H is the flux density of the permanent magnets and b is a constant. This torque will cause the coil to rotate. The rotation is resisted by the spring with a torque of $k\theta$.

Friction in the pivots causes a damping torque proportional to the angular velocity, $c(d\theta/dt)$. In addition, eddy currents set up in the coil frame also produce a damping torque of $dH^2(d\theta/dt)$, where d is a constant. The sum of these torques is equal to the moment of inertia of the coil and pointer times the angular acceleration

$$bHi - k\theta - c\frac{d\theta}{dt} - dH^2\frac{d\theta}{dt} = I\frac{d^2\theta}{dt^2} \qquad (5\text{-}38)$$

Thus the motion of the pointer is given by

$$\frac{d^2\theta}{dt^2} + \frac{c + dH^2}{I}\frac{d\theta}{dt} + \frac{k}{I}\theta = \frac{bH}{I}i \qquad (5\text{-}39)$$

so that the galvanometer is a second-order instrument with

$$\omega_n = \sqrt{\frac{k}{I}} \qquad \rho = \frac{c + dH^2}{2\sqrt{kI}} \qquad K = \frac{bH}{k} \qquad (5\text{-}40)$$

Other examples of second-order instruments are accelerometers, pressure transducers, and U-tube manometers.

The damping factor is a very important parameter in the response of a second-order instrument. The characteristic equation for the homogeneous part of Eq. 5-36 is

$$s^2 + 2\rho\omega_n s + \omega_n^2 = 0 \tag{5-41}$$

which has the roots

$$s = -\rho\omega_n \pm \omega_n \sqrt{\rho^2 - 1} \tag{5-42}$$

The nature of the response will be determined by the value of the damping factor. If ρ is greater than 1, both roots will be real. The homogeneous solution will then be

$$y_h = e^{-\rho\omega_n t} \, [A \, \sinh \, (\omega_n \sqrt{\rho^2 - 1})t + B \, \cosh \, (\omega_n \sqrt{\rho^2 - 1})t] \tag{5-43}$$

This nonoscillatory motion dies out exponentially and is called *overdamped.*

The critical value of the damping factor is one. For $\rho = 1$ there are two repeated roots of the characteristic equation. The homogeneous solution is then

$$y_h = e^{-\omega_n t}(A + Bt) \tag{5-44}$$

This *critically damped* motion is also nonoscillatory. When $\rho < 1$, the *underdamped* homogeneous solution is

$$y_h = e^{-\rho\omega_n t} \, [A \, \sin \, (\omega_n \sqrt{1 - \rho^2})t + B \, \cos \, (\omega_n \sqrt{1 - \rho^2})t] \tag{5-45}$$

which can also be written

$$y_h = Y e^{-\rho\omega_n t} \, \sin \, [(\omega_n \sqrt{1 - \rho^2})t + \phi] \tag{5-46}$$

These homogeneous solutions add to the particular solution for the input to the instrument to give the instrument's dynamic response.

Step Response

The particular solution for a step input is

$$y_p = Kx_0 \tag{5-47}$$

For the overdamped case, the step response is

$$y = Kx_0 \left\{ 1 - e^{-\rho\omega_n t} \left[\cosh (\omega_n\sqrt{\rho^2 - 1})t + \frac{\rho}{\sqrt{\rho^2 - 1}} \sinh (\omega_n\sqrt{\rho^2 - 1})t \right] \right\}$$

(5-48)

When $\rho = 1$, we find

$$y = Kx_0 [1 - e^{-\omega_n t}(1 + \omega_n t)]$$

(5-49)

and with underdamping,

$$y = Kx_0 \left(1 - e^{-\rho\omega_n t} \left\{ \frac{1}{\sqrt{1 - \rho^2}} \sin [(\omega_n\sqrt{1 - \rho^2})t + \phi] \right\} \right)$$

(5-50)

where

$$\phi = \sin^{-1} (\sqrt{1 - \rho^2})$$

(5-51)

These solutions are shown in Fig. 5-17.

First we note that $1/\rho\omega_n$ now plays the part of a time constant. For small values of $1/\rho\omega_n$ the response quickly approaches the static value, whereas large values cause a slow response. In addition, the value of ρ also determines whether the response *overshoots* the static value. The step response always overshoots the static value for damping factors less than 1.

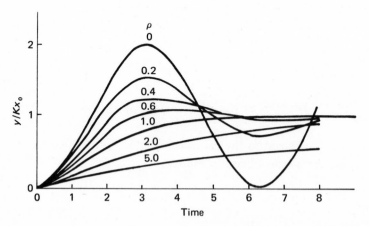

Figure 5-17 The response of a second-order system to a step input as a function of several damping factors.

Since we usually want an instrument to respond quickly to a step input, we would ordinarily select an underdamped instrument in order to reduce the time constant. At first glance, we might choose a critically damped instrument since it would have the quickest response without overshooting the static value. However, if we allow some overshoot, the time required to reach the true static value will be reduced. If an overshoot of 5% is allowed, a damping factor of 0.69 will result in a response that comes to within 5% of the static value in about half the time of a critically damped instrument. *Thus, most instruments are designed with damping factors near 0.7.* We also note that the response will be faster if the natural frequency ω_n is increased.

Ramp Response

Both the ramp input and the sinusoidal input to an instrument are continually changing. The homogeneous part of the solution dies out fairly quickly from the damping, so that eventually only the particular solution is left. Since the particular solution persists, it is called the *steady-state response.* For constantly changing inputs, such as the ramp input, the steady-state response is of primary interest. For the ramp input of Eq. 5-7, the steady-state response is

$$ y = Ka\left(t - \frac{2\rho}{\omega_n}\right) \tag{5-52} $$

Again, there is a velocity error in the response. The response lags behind the input by $2\rho/\omega_n$. To reduce the velocity error, we need low values of the damping factor and high natural frequencies. With low values of ρ, the output will oscillate when the signal is first applied, but this oscillation will quickly die out if ω_n is large.

Frequency Response

For a sinusoidal input, the steady-state response of a second-order instrument is

$$ y = \frac{KX}{[(1 - \omega^2/\omega_n^2)^2 + (2\rho\omega/\omega_n)^2]^{1/2}} \, [\sin{(\omega t + \phi)}] \tag{5-53} $$

with

$$ \phi = \tan^{-1}\left(-\frac{2\rho\omega/\omega_n}{1 - \omega^2/\omega_n^2}\right) \tag{5-54} $$

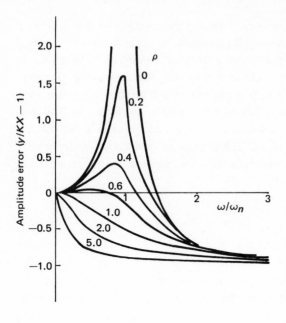

Figure 5-18 The amplitude error of a second-order instrument in response to a sinusoidal change in input as a function of frequency ratio and damping factor.

The frequency response involves both an *amplitude error* and a *phase error*. These errors are plotted in Figs. 5-18 and 5-19.

For $\rho < 1$ the amplitude of the response has a peak at

$$\omega = \omega_n\sqrt{1 - 2\rho^2} \tag{5-55}$$

This peak is called *resonance* and for zero damping occurs at the natural frequency, producing a response that grows linearly with time. For nonzero

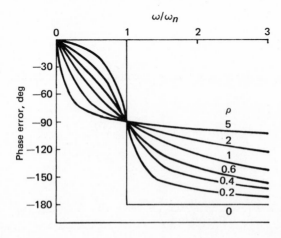

Figure 5-19 Phase error of a second-order instrument having a sinusoidal input as a function of frequency ratio and damping factor.

damping, the amplitude at resonance is finite. The phase error is always $90°$ at $\omega = \omega_n$ and approaches $180°$ as the frequency ratio ω/ω_n gets very large.

The graphs in Figs. 5-18 and 5-19 show the most desirable value of the damping factor. For ρ near 0.65, the amplitude curve is nearly flat for a wide band of frequencies. The phase angle curve is also nearly linear with respect to frequency for this value of ρ. A linearly varying phase error is desirable because this will produce minimum distortion of the wave shape.

Optimum response of a second-order instrument to inputs of varying frequencies is obtained when the damping factor is near 0.65. Since $\rho = 0.69$ also gives the best step response, second-order instruments are almost always designed to have a damping factor in the range 0.65–0.70.

The frequency response can also be used to calculate the dynamic loading effect of the instrument. The power absorbed by the instrument will depend on the nature of the input. If the input to the instrument is an oscillatory displacement, then the mean power consumed will be

$$P = \rho \omega_n \omega^2 \delta^2 \qquad (5\text{-}56)$$

and the power input from an oscillating force is

$$P = \frac{F^2 \sin^2 \phi}{4\omega_n \rho} \qquad (5\text{-}57)$$

Since the phase angle is $90°$ at $\omega = \omega_n$, Eq. 5-57 shows that the maximum loading error occurs when the input frequency is ω_n. The second-order instrument has a variable input impedance with the input impedance determined by the input frequency ω.

From the above results, we see that displacement measurements should be made using an instrument having a low natural frequency, whereas force or acceleration measurements should be made by an instrument with high ω_n. Seismometers are, in fact, designed with ω_n in the range 2–5 Hz, whereas accelerometers normally have a natural frequency of 500 Hz or more. A similar calculation for a first-order instrument will show that dynamic loading errors in first-order instruments are always reduced by making the time constant smaller.

Example 5-3 A pressure transducer has a natural frequency of 80 Hz, a damping factor of 0.7, and a static gain of 8 mV/psi. A step input of 100 psi is applied. What is the error in the response at 0.1, 0.5, and 1.0 s?

Solution Since $\rho < 1$, the response is given by Eq. 5-50. The natural frequency is

$$\omega_n = \frac{80}{2\pi} = 12.72 \text{ rad/s}$$

Since the damping factor is 0.7,

$$\sqrt{1 - \rho^2} = 0.713$$

and the phase angle

$$\phi = \sin^{-1} \sqrt{1 - \rho^2} = 0.794 \text{ rad } (45.5°)$$

Then

$$\rho\omega_n = 8.91\sqrt{1 - \rho^2}\,\omega_n = 9.07$$

Thus the response is

$$y = 800[1 - 1.4e^{-8.91t} \sin (9.07t + 0.794)]$$

t	$e^{-8.91t}$	$\sin (9.07t + 0.794)$	y (mV)	Error (%)
0.1	0.41	0.991	344.8	−56.9
0.5	0.0115	−0.809	810.4	+1.3
1.0	1.32×10^{-4}	−0.358	800.05	+0.08

Example 5-4 A machine has a suspected vibration at a frequency of 600 Hz. Two accelerometers are available to measure the vibration. One is a strain gauge device with a natural frequency of 800 Hz and a damping factor of 0.6. The other is a solid-state device with $\omega_n = 10,000$ Hz and $\rho = 0.06$. What are the steady-state errors for each device?

Solution The amplitude of the response is given by Eq. 5-53,

$$\frac{y}{Kx} = \frac{1}{[(1 - \omega^2/\omega_n^2)^2 + (2\rho\omega/\omega_n)^2]^{1/2}}$$

and the phase error by Eq. 5-54. For the strain gauge device,

$$\frac{\omega}{\omega_n} = \frac{600}{800} = 0.75$$

$$\rho = 0.6$$

Thus

$$\frac{y}{Kx} = \frac{1}{[(1 - 0.75^2)^2 + (2 \times 0.6 \times 0.75)^2]^{1/2}} = 0.9994$$

The amplitude error is

$$\left(\frac{y}{Kx} - 1\right) \times 100 = -0.06\%$$

Also

$$\phi = \tan^{-1}\left(\frac{2 \times 0.6 \times 0.75}{1 - 0.75^2}\right) = \tan^{-1}(-2.055)$$

so the phase error is

$$\phi = -64°$$

For the solid-state device,

$$\frac{\omega}{\omega_n} = \frac{600}{10,000} = 0.06$$

$$\rho = 0.06$$

Then

$$\frac{y}{KX} = 1.00357$$

for an amplitude error of 0.36%. The phase error is

$$\phi = -0.41°$$

Although the strain gauge accelerometer has a smaller amplitude error, the amplitude errors of both devices are acceptable. Thus, the solid-state device should be used because it has practically no phase error.

5-6 HIGHER ORDER INSTRUMENTS

Most basic transducers and instruments are second-order or less. However, when two or more instruments are used together, the overall response will be that of a *higher order instrument.* For example, if the output of an anemometer, a first-order transducer, is recorded using an oscillograph, a second-order instrument, the response of the measurement system will be described by a third-order equation.

In general, there is little that we can say about the response characteristics of higher order instruments. Each case must be analyzed individually to determine the nature of the response and the effect of the system parameters.

When a higher order system can be broken down into first- and second-order components, as in our anemometer–oscillograph example, some conclusions can be drawn about the frequency response.

Let us consider an instrument system consisting of a first-order instrument connected to a second-order instrument. The first-order component will have a time constant of τ and a static gain K_1. The second-order component has a natural frequency ω_n, a damping factor ρ, and static gain K_2. If the input to the first-order part of the system is

$$x = X \sin \omega t \qquad (5\text{-}58)$$

the output will be

$$y_1 = \frac{K_1 X}{(1 + \omega^2 \tau^2)^{1/2}} \sin (\omega t + \phi_1) \qquad (5\text{-}59)$$

with

$$\phi_1 = \tan^{-1} (-\omega \tau) \qquad (5\text{-}60)$$

This is now the input to the second-order part of the system. Using this input, we find that the steady-state response of the second-order component, which is the final output of the system, is

$$y = \frac{K_1 K_2 X}{(1 + \omega^2 \tau^2)^{1/2} [(1 - \omega^2/\omega_n^2)^2 + (2\rho\omega/\omega_n)^2]^{1/2}} \sin (\omega t + \phi_1 + \phi_2)$$

$$(5\text{-}61)$$

where

$$\phi_2 = \tan^{-1}\left[-\frac{2\rho\omega/\omega_n}{1-(\omega/\omega_n)^2}\right] \tag{5-62}$$

We see that the amplitudes of the two responses have multiplied, while the phase errors have added. This result can be extended to any instrument system made up of zero-, first-, and second-order components. The ratio of output to input amplitude for the system is the product of the amplitude ratios of the components, whereas the phase error of the system is the sum of the phase errors of the components.

Example 5-5 An oscillograph with $\omega_n = 3600$ Hz, $\rho = 0.65$, and $K = 0.5$ mm/mV records the output of a pressure transducer that has $\omega_n = 80$ Hz, $\rho = 0.45$, and $K = 10$ mV/kPa. How will this system respond to a 20-kPa pressure oscillating at 120 Hz?

Solution For the transducer,

$$\frac{\omega}{\omega_n} = \frac{210}{80} = 1.5 \qquad \rho = 0.45$$

Thus,

$$\frac{y}{x} = \frac{10}{[(1-1.5^2)^2 + (2 \times 0.45 \times 1.5)^2]^{1/2}} = 5.44 \text{ mV/kPa}$$

and

$$\phi_1 = \tan^{-1}\left(-\frac{2 \times 0.45 \times 1.5}{1-1.5^2}\right) = -132.8°$$

For the recorder,

$$\frac{\omega}{\omega_n} = \frac{120}{3600} = 0.033 \qquad \rho = 0.65$$

and

$$\frac{y}{x} = \frac{0.05}{[(1-0.33^2)^2 + (2 \times 0.65 \times 0.033)^2]^{1/2}} = 0.499 \text{ mm/mV}$$

$$\phi_2 = \tan^{-1}\left(-\frac{2 \times 0.65 \times 0.033}{1 - 0.033^2}\right) = -2.5°$$

Therefore, for the system,

$$\frac{y}{x} = 5.44(0.499) = 2.715 \text{ mm/kPa}$$

$$\phi = -132.8 - 2.5 = -135.3°$$

and the system has an output of 54.3 mm. Thus, there is an amplitude error of -45.7% and a phase error of $-135.3°$ in the oscillograph output.

5-7 SUMMARY

Accurate engineering measurements require careful attention to the loading effects and dynamic response of the test instruments in addition to the accuracy and precision requirements. The input impedance, either electrical or mechanical, of the instrument must be properly matched to the output impedance of the test apparatus to avoid loading errors caused by power transfer to the instrument. Either high or low input impedance relative to the output impedance may be needed, depending on the nature of the quantity being measured.

When selecting an instrument system, one should look for certain characteristics that will provide good dynamic response. First-order instruments should have a time constant as small as possible. Second-order instruments should have a damping factor in the range 0.6-0.7. Input impedance considerations determine whether high or low natural frequency is desirable. High input impedance is most common. This requires a high natural frequency for the instrument relative to the measurement frequency. Normally the instrument is selected so that the input frequency will be no more than 40% of the instrument natural frequency.

As a practical matter, the experimenter often has little or no control over the characteristics of the instruments. The time constant of the most common first-order instrument, the thermometer, varies with the type, velocity, and even the temperature of the fluid whose temperature is being measured. Since these time constants can be rather large—30 s to 1 min for many thermometers—the test plans and procedures must be adjusted to compensate for the response errors.

PROBLEMS

5-1 Show that a force measurement should be made with a device having a high spring rate, whereas deflection measurements should be made with low-spring-rate devices.

5-2 If a voltage-generating device with an output impedance of R_i is connected to a voltmeter with an input impedance of R_m, the voltage measured by the voltmeter will be

$$E_m = \frac{R_m}{R_m + R_i}E$$

where E is the voltage generated by the device with an open output circuit. Since the power consumed by the meter is E_m^2/R_m, show that the maximum loading error occurs for $R_m = R_i$.

5-3 Displacement measurements are often made using the slide-wire potentiometer shown in the accompanying figure, where the displacement is determined from the unknown resistance R. If the input impedance of the voltmeter is high enough, the unknown resistance R is determined from

$$R = R_0 \frac{E_m}{E_0}$$

What is the correct expression for R if the meter resistance is low enough to cause a loading error?

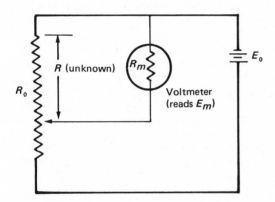

5-4 What should be the input impedance of the voltmeter in Fig. 5-4 if the error in E_m is to be no larger than 1%?

5-5 Newton's law of cooling is

$$q = hA\,(T_f - T_w)$$

where q is the heat-transfer rate (W/m^2), h is the film coefficient (W/m$^2 \cdot$ °C), A is the surface area (m^2), T_f is the fluid temperature (°C), and T_w is the wall temperature (°C). Using this equation, find the time constant of a thermometer in terms of the physical parameters.

5-6 A thermocouple is initially at 70°F and is to be used to measure the temperature of an oil bath whose temperature is *about* 170°F. If the time constant of the thermocouple is 10 s, how long should you wait after dunking the thermocouple in the bath before reading it to be sure that the indicated temperature is within $5°F$ of the true temperature?

5-7 A resistance thermometer has a resistance of 100 Ω at $20°C$ and a sensitivity of 0.025 $\Omega/°C$. It is placed in an oven at $300°C$. After 8 s have elapsed, the resistance is 104 Ω. Estimate the time constant of the thermometer.

5-8 An anemometer has a time constant of 0.4 s. If this instrument is used to measure turbulent winds, what is the highest frequency of fluctuation that can be measured if the amplitude error is to be less than 5%?

5-9 Show that a U-tube manometer is a second-order instrument with

$$\omega_n = \sqrt{\frac{g}{L}} \qquad \rho = \frac{16\mu}{\rho_f D^2} \sqrt{\frac{L}{g}} \qquad K = \frac{1}{\rho_f g}$$

where L is the length of the fluid column, D is the diameter of the tube, ρ_f is the fluid density, μ is its viscosity, and g is the acceleration due to gravity.

5-10 A step input is made to a critically damped ($\rho = 1$) second-order instrument. Make a graph showing the time required for the response to reach 95% of the true value as a function of the natural frequency of the instrument.

5-11 Calculate the frequency ratio ω/ω_n for which the amplitude error of a second-order instrument is -1% when $\rho = 0.7$.

5-12 For $\omega < \omega_n$, show that the maximum amplitude error occurs at $\omega/\omega_n = \sqrt{1 - 2\rho^2}$ for $\rho < 0.707$ and at $\omega = \omega_n$ for $\rho > 0.707$.

5-13 A seismometer has a natural frequency of 6 Hz and $\rho = 0.35$. What will be the amplitude error for an input frequency of 7 Hz?

5-14 An oscillograph has a natural frequency of 3500 Hz and a damping factor of 0.65. What is the range of frequencies that can be recorded by this instrument so that the amplitude error will be less than 5%?

5-15 The output of a pressure transducer is a 50-Hz sine wave with an amplitude of 400 mV. The specification sheet for the transducer gives the following data: $\omega_n = 30$ Hz, $\rho = 0.1$, $K = 8$ mV/psi. What is the amplitude of the pressure wave being measured?

5-16 The output of an anemometer with a time constant of 1 s is recorded on an oscillograph with a natural frequency of 400 Hz and a damping factor of 0.69. Plot the frequency response (amplitude ratio and phase error) of the system for input frequencies from 0 to 160 Hz.

5-17 A strip-chart recorder can be considered to be two first-order instruments in series. The time constants for the two components of the recorder are 0.001 s and 0.25 s. The recorder is connected to a thermocouple with a time constant of 3 s. What are the amplitude and phase errors of the system when the temperature is varying with a frequency of 2 Hz?

BIBLIOGRAPHY

Beckwith, T. G., and N. L. Buck: *Mechanical Measurements,* Addison-Wesley, Reading, Mass., 1969.

Doebelin, E. O.: *Measurement Systems Application and Design,* McGraw-Hill, New York, 1975.

Graham, A. R.: *An Introduction to Engineering Measurements,* Prentice-Hall, Englewood
    Cliffs, N.J., 1975.
Holman, J. P. *Experimental Methods for Engineers,* New York, McGraw-Hill, 1971.
Phelan, R. M.: *Dynamics of Machinery,* McGraw-Hill, New York, 1967.
Prentis, J. M.: *Dynamics of Mechanical Systems,* Longman, London, 1970.
Thomson, W. T.: *Theory of Vibration with Applications,* Prentice-Hall, Englewood Cliffs,
    N.J., 1972.

TEST SEQUENCE
AND EXPERIMENTAL PLANS

Having completed the instrumentation of a test and checked the apparatus for accuracy and response, we then examined the possibility of variable reduction using the powerful tool of dimensional analysis. At this point the experimenter may be ready, even eager, to "turn on the juice" or otherwise set the test in motion. Impatience and hasty starts will seldom bring quick completion. Rather, haste may result in overloading or straining some test component, and, if it does not produce physical trouble, it is sure to result in an inefficient test. Overlapping and duplication may occur over some ranges of the apparatus and incomplete coverage in others. The control over known and computed variations of the surroundings may be sloppy, and any accounting of natural or extraneous variations is likely to be nonexistent. Months later the engineer may realize that the apparent, even obvious, effect of high velocity was in truth due to the chance running of high-velocity points on second shift, and that the experiment inextricably mingled a test of the apparatus with a test of the operating personnel, or weather variations, or some regular defect in the instruments, and so on.

In this chapter we shall see how the engineering experiment can be rationally planned, point by point, to give speed of testing, minimization of error, maximization of useful data, and maximum control of extraneous and outside influences. In short, we shall inquire into the ways in which an experiment can be made efficient without loss in meaning or accuracy.

6-1 PLANNING FOR BALANCE CHECKS

In Example 3-4 and Fig. 3-4 we discussed a heat-transfer experiment in which an *energy balance* was applied to the energy leaving the hot fluid and entering the cold fluid. In that particular example, the balance was in doubt because of the accuracy and/or precision of the temperature measurements and the very small rise in the cold water temperature. That example suggests two precautions in all those tests in which some form of balance exists: First, the planner should always have enough instruments to allow a balance check to be made; and second, the test conditions and instrument accuracy should be matched so that the comparison serves as a true indication of proper test operation. Many tests can be, and are, carried out without a balance check, but failure to utilize such a check is a sure sign of a sloppy setup or an investigator too willing to depend on "luck" or "faith."

There are, of course, tests in which obtaining the balance terms may involve more difficult measurements than operating the test without this check, but it is often true that adding one or two inexpensive instruments allows the test planner to be sure that the total data set is self-consistent.

Some common tests in which balance equations might be useful are the following:

1. *Energy flow in heat-transfer equipment.* Here it is usually possible to measure the energy given up by the hot fluid and the amount gained by the cold, since such devices are always constructed so that very little energy is lost to the surroundings. The exception occurs when one fluid is condensing or boiling. In this type of exchanger, the enthalpy loss or gain can be computed only when the actual weight per unit time of fluid boiled or condensed is found, a measurement that is notoriously difficult to make in many cases.

2. *Current flow in electrical networks.* This is really a mass balance, since we are assuming that the same number of electrons that enter a network will ultimately leave it. The range of application of this conservation law is very large and covers such systems as simple Kirchhoff networks, vacuum tube and transistor tests, three-phase problems, and so on. The main precaution is that the analyst must fully understand the phase relationships in the system and be sure that the ammeters are measuring the proper current at whatever point they are inserted.

3. *General mass balance in fluid systems.* This is so basic a consideration that it appears trivial; yet it is often overlooked in the design of test systems. Whenever we place two or more flow-metering devices in the same line, such that one will check the other, we are making use of the basic continuity principle that mass is conserved. This is often done to calibrate an air-flow orifice against a

standard nozzle or to calibrate a fluid-flow meter against a weigh-barrel measurement. If it can be done conveniently, permanent installation of two different flow meters in the same line is far better.

4. *Momentum exchange in conservative fluid systems.* The use of the momentum balance to check for error is often a rather subtle problem, because momentum is so easily lost in exchange situations through turbulence, friction, and nonelastic collisions. In some tests involving flowing fluids, the investigator can be sure that friction plays a small part so that a momentum balance is possible. In a jet engine tailpipe, for example, radial and linear velocities will both be present, and momentum will be interchanged between these two forms as the gas travels down the duct. A series of measurements should show the summed momentum to be constant if all is well. Probably the most powerful application of the momentum balance occurs in the study of nuclear tracks on cloud chamber photos.

There is actually a vast variety of balance equations available to investigators. Charge, voltage, or field strength might be conserved in electrical tests. Enthalpy, free energy, or entropy could be conserved in thermal or chemical systems. Head, pressure, or specific energy might be balanced in certain fluid experiments. When planning a test, the investigator should always consider whether a conservation equation exists in the test and if it is readily checked by measurement.

In many common tests a balance cannot be applied for various reasons. Some examples follow.

1. *Energy balance on an engine, a fan, or a compressor.* All such devices essentially transform one form of energy (thermal, electrical, mechanical) into another form (usually mechanical). Interposed between the numerical output and the numerical input is the conversion efficiency, a number usually unknown until basic testing work is complete. Thus we can say that the output cannot exceed the input, but we can say no more than this without knowing a great deal about the device.

2. *Combustion tests.* Although it is theoretically possible to estimate the stack loss, insulation losses, and so on, of a combustion system and then to estimate the energy release of the fuel through measurement of the flue products, the actual energy balance is of such questionable precision that it can seldom be used with any confidence to check other measurements.

3. *Mass balance in large fluid systems.* In large fluid-flow tests involving whole rivers, large duct work carrying low-pressure air, and similar cases, it is quite difficult to arrive at mass balances at two different stations that will agree with one another. The main reason is that local velocities across a river or in a duct may vary radically from point to point, so that very many careful velocity checks must be made at a variety of locations at a section. When a temperature

gradient is also imposed on a velocity profile, with energy of the stream to be computed, the problem is compounded.

The use of a balance check in testing work is often overlooked or ignored by engineers simply because they do not see any great use for it. In problems involving a difficult flow measurement, such as liquid–metal flow in a closed system, the engineer may reject the use of two flow meters, saying, "Which one should I use? Should I average them?" Both might be averaged if they read close to each other without one being consistently high or low. If they disagree by 20 or 30%, neither should be used, at least until it is determined just what is the matter. If the measurement is accurate but very imprecise, one might reject all data that did not show a flow balance to within, say, 5%.

Example 6-1 A weigh barrel measures 15 lb of water ($\rho = 62.5$ lb/ft^3) in 1 min, while a volume flow meter in the same line registers 0.27 ft^3/min. What is the amount of imbalance, and what is the percent error, taking the weigh-barrel measurement as correct?

Solution The conservation (of mass) equation in this case is

$$\text{Time} \times \text{weight} = \text{volume flow} \times \text{density}$$

or

$$1 \times 15 \overset{?}{=} 0.27 \times 62.5$$
$$15 \neq 16.9$$

So the imbalance is $16.9 - 15.0$, or 1.9 lb/min, and the percent error is $1.9/15 = 12.7\%$.

6-2 SPACING OF TEST POINTS

In very few tests is it possible to estimate exactly the correct amount of testing. Too little testing and the law or function may not be found, the accuracy of constants may be poor, or some small effect of large theoretical importance may be missed. Too many data, on the other hand, and the test becomes overlong, the data processing endless and expensive, and even the presentation of material becomes more difficult. There are some tests in which excessive numbers of data may actually obscure rather than reveal certain important effects. For example, a stress-versus-strain test of a single tensile strength specimen might easily show the little dip region at the end of the elastic range found with some low-carbon steels. But suppose that we tested half a dozen specimens, in each of which the dip region might be displaced slightly along the stress or strain axis when compared with the others. If all the points are plotted in an undifferentiated manner, the dip region could disappear completely, as shown in Fig. 6-1. The multispecimen plot is best if we wish to determine the average tensile strength properties of the steel, but it is not as good as the single-specimen plot if we are studying the general behavior of metals in tension.

The most obvious way in which an experimental plan can be made compact and efficient is to space the variables in a predetermined manner. If we are searching for a functional relationship between an independent variable X and a dependent variable Y, we hope that the function can be represented by a curve or line on (X, Y) coordinates. Such a line is made up of an infinite number of separate points, and we must choose from this infinite family some finite and practical number to represent the function. If the function has two independent variables, a complete data population will fill an entire plane area, and so on.

Thus, the choice of the finite population of test data is a legitimate and important area of pretest planning, although it can be done, and unhappily often is, while the test is under way. The most obvious way to start test-point selection is to decide on the end points or limits of the test apparatus, which will give the *test envelope* that encloses the complete family of data. Some typical limitations on engineering test equipment are compressor surge line, metallurgical temperature limits, structural-rotational speed limits, power-handling limits imposed by

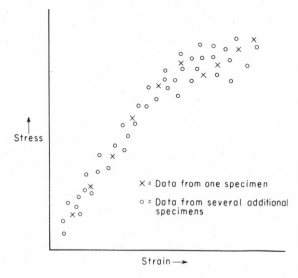

x = Data from one specimen

o = Data from several additional specimens

Figure 6-1 Sketch showing how the "dip region" in a stress–strain curve for steel might be obscured if too many data from several different tests are plotted together.

dynamometer size, thermal-input limitations caused by furnace size, flow limitations imposed by pipe areas, and so on. Such limit points can often be found or computed before operation, although often they must be checked by test, particularly when untried items are undergoing initial performance checks.

For the simple XY function, two points enclose all others. An XYZ function envelope is a plane area, or *map*, and a large number of points may be needed simply to outline its extent. Functions involving more variables than this are usually broken down to a series of maps.

Let us consider the criteria for spacing points in our envelope, taking the basic XY function. (Note that the XYZ function is reducible to a set of XY functions as illustrated in Fig. 6-2.) There are two major criteria governing test-point selection.

1. *Relative accuracy of data in different regions of the test envelope.* This important criterion, so often overlooked by engineers, stems directly from our discussion of error and uncertainty in Chaps. 2 and 3. Many tests will show data that have unequal precision percent error over the entire test envelope (see Example 3-1). In many mechanical engineering tests, we suspect that low-power, low-head runs will be the most imprecise. Civil engineers always doubt their low-head, low-flow points the most, whereas in certain electrical measurements very high resistance readings are questionable because of "sneak circuits." When our error analysis shows us that a part of the test envelope is in greatest question, we should naturally fill in that portion with more than its normal share of points. We can make no fixed rule on how many additional points to take in the doubtful regions, but the general rule of Eq. 2-25 relating precision improvement to additional readings can be expected to hold, so that four points are twice as effective as one; and nine, three times as effective. In Example 3-1, where the precision percent error at one end of the envelope was found to be more than 2.5 times better than at the other, we should theoretically take six to seven times as many points at the bad end to make the curve equally precise

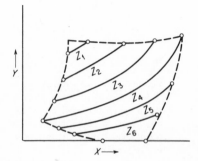

Figure 6-2 A possible XYZ envelope with Z as parameter. The dashed lines enclose the total family of points that can be attained.

over its length. This is practical nonsense, and the experimenter must use judgment.

2. *Nature of the experimental function.* In the majority of engineering experiments, the investigator usually has a good idea of the experimental function. If an uncertainty criterion does not suggest the point spacing and the function is partially or completely known, it is often well worth the time to set up a solution to find an even-spacing plan. For example, the loss of pressure through a new fitting as a function of flow and density may be unknown until tests are carried out, but the experimenter familiar with hydraulic work should certainly suspect a function of the type

$$\Delta P = k\rho \frac{V^2}{2g} \tag{6-1}$$

where k may be constant with changing flow velocity V or might be a weak function of V.

If the function is expected to be such a simple one, it can often be transformed to a linear form by one or another kind of algebraic transformation. The pressure loss-versus-velocity function becomes

$$\ln \Delta P = \ln \frac{k\rho}{2g} + 2 \ln V$$

If we are willing to accept a plot of $\ln \Delta P$ versus $\ln V$, we can have a correctly spaced curve by taking equal increments of $\ln V$ instead of V alone. Furthermore, we do not have to know the value of k, g, or ρ, or know that the exponent of V is 2.

Some other examples of similar transformations are

$$Y = \frac{A + B}{X^2}$$

where correct spacing occurs if we take equal increments of $1/X^2$ rather than X alone and plot $1/X^2$ versus Y,

$$Y = Ae^{-bX}$$

which becomes

$$\ln Y = \ln A - bX$$

and we should plot X versus ln Y taking equal increments of X,

$$Y = A \ln BX$$

or

$$Y = A(\ln B + \ln X)$$

And we see that Y should plot versus ln X, and ln X should take equal intervals.

It must be kept in mind that we are not adjusting the spacing of test points simply to get a "symmetrical" or "pleasing" curve. There is, in fact, one basic reason for considering point spacing at all. It is our desire to have every part of our experimental curve or map have the same precision as every other part. We may or may not be able to achieve this ideal in any given test. We should, however, never fail to try to bring it about.

6-3 SEQUENCE OF EXPERIMENTAL TESTING

Having spaced the test points through a consideration of precision, we still do not know just what sequence we should follow in putting our apparatus into these chosen configurations. There are many types of experiments in which little or no choice exists as to the sequence of operation. In astronomy and many of the so-called earth sciences, outside factors force a time and sequence on the investigator. We must observe Mars, not when we might wish, but when the weather is clear, the planet close, and the moon not bright. The usual sequence of seasons may not be particularly convenient for running some planting experiments, but we cannot change them.

In engineering, such situations are somewhat unusual. More common is the experiment that we shall call *irreversible*. This is a test that proceeds irrevocably from past to future without chance of alteration. Most obvious of this class are those tests involving endurance under extreme conditions, where the test item suffers continuous and progressive deterioration. Many materials-testing experiments are irreversible. Suppose that we plan a test on a steel tensile specimen and decide to apply our preselected load values at random. We apply 6000, then 1000, then 900, then 15,000 lb, and so on. Such a plan is certainly defective. The first load that we apply above the elastic limit will permanently deform the piece, and all subsequent readings will be made on a deformed specimen. Other examples of irreversible tests are all tests in which chemical changes (such as corrosion) may occur, all tests involving metal fatigue where this fatigue is significant to the operation of the test, and all tests in which high temperatures, radiation fields, or high-gravity forces are progressively changing the crystalline structure of the test piece.

It could be argued that all tests are basically irreversible in the sense that no piece of apparatus ever returns exactly to its original configuration after use. Usually, the changes wrought by testing are so small as to be below the level of detection, and we say that such tests are reversible and maintain that the apparatus can be returned at will to any previous configuration. All such tests admit a choice of point sequence, of which we shall discuss two basic kinds. We may start with an independent variable at its upper or lower extreme value and change it in steps until the other extreme is reached. Or we may run the selected points in a perfectly random fashion, now high, now low. The first plan we shall call the *sequential plan* and the second, the *random plan.* The fact that the sequential plan is now followed in almost all engineering tests is remarkable, because the random plan makes more sense in most reversible experiments.

The sequential plan is obviously essential for irreversible tests of the materials type. There are other, more subtle tests in which sequential plans are also desirable. The best example and one familiar to every young engineer is the classic pipe friction experiment. At first glance it might not be evident why a sequential plan is desired in this case. As it happens, a laminarly flowing fluid, if the Reynolds number is slowly and carefully increased, tends to remain in laminar motion well into the transition region, whereas the opposite (continued turbulence) occurs as the Reynolds number is shifted from high to low values. Figure 6-3 shows the data of a group of college juniors who were told that this effect was possible and could be noted by running in ascending and then descending sequence of Reynolds numbers. Had Reynolds numbers been selected in a random manner, now laminar, now turbulent, it is doubtful that this small effect would have been found. This is an experiment in which *sequence itself is a parameter of the test.* Other simple tests that show similar behavior are any test of an iron core inductor in which the hysteresis pattern may depend on a previous test point and tests of friction in which transitions are made from starting to sliding friction and back.

The great majority of tests in engineering are best handled by a partially or completely randomized plan. The arguments for such plans are quite convincing, and a few will be advanced at this point.

Natural effects may show a general trend during a test series. The barometric pressure may increase; the surrounding temperatures may rise or sink slowly; the humidity may change. If the controlled variable X is also changing in a regular manner, the dependent variable R may show the effect of both X variation and weather variation. If X is varied in a random

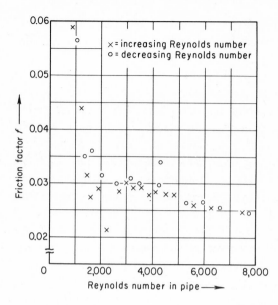

Figure 6-3 A plot of friction factor versus Reynolds number for the flow of water in copper tubing. The test was run with N_{re} increasing continuously and then decreasing continuously, with the two kinds of points distinguished. Note that in the range 1500–5000 the decreasing points show higher f values than the increasing ones, suggesting a sequence effect on the test.

manner, there can be no chance of mistaking weather-induced trends for an X effect.

Human activities may show a trend during the test series. Most obvious is increasing skill or, alternatively, increasing boredom among data takers and operating personnel. Are certain effects caused by increasing X variable, or sloppy reading as shift change approaches? There need be no confusion about such factors if the X variable is varied in a random manner.

Mechanical effects may produce a trend with regular X variation. This is probably the most important reason for using a random plan. Suppose that we have a control, instrument, gauge, or other device that is sticky. If the previous reading on the apparatus is a high one, the instrument will stick high, whereas if a low reading precedes, the device will stick low. Now, if we go from low to high in regular, sequential steps, what is the effect? Every single reading, except perhaps the first one, will be low, and the entire test will have a regular and hard-to-spot bias. Suppose instead that we randomize our point selection so that there are just about as many points approached from above as from below. The data may show some scatter, *but they will scatter around the correct values.* Figure 6-4 shows such a situation. This occurs in test work in a variety of guises. All sorts of thermal and engine tests will show errors from a failure to reach steady state. If a sequential plan is used, the data will always be on the same side of steady-state conditions (too hot or too cold), and regular errors will result. Dirty manometer tubes behave like sticky instruments, as do orifice lines that are

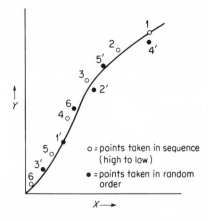

o = points taken in sequence (high to low)

• = points taken in random order

Figure 6-4 Plot illustrating the effect of a "sticky" instrument. In this case, the reading is "high" if it follows a higher reading and "low" if it follows a lower reading. Note that the random set of points will bracket the "true" curve, whereas the sequential plan yields a curve that is misleadingly high.

partially blocked or pressure lines with small leaks. Unless the experimenter is very familiar with the system, he or she risks regular and troublesome error by following a sequential plan when it is not necessary.

Perhaps it is not putting the matter too strongly to say that the *only* justifications for the sequential plan are (1) an experiment known to be either irreversible or having some aspect that can only be shown by taking data in some regular sequence; or (2) an experiment of such length, cost, or difficulty that randomization is not practical. An example of this latter case might be a nuclear reactor with a time to reach thermal steady state that is on the order of days, so that each change in operating conditions must be as small as possible.

In the next sections, we shall see how partial and complete randomization can be achieved through special "block" plans. For many experiments, randomization is best obtained by some simple "gaming" method. For example, the chosen runs can be numbered and the numbers drawn from a hat as in a lottery. If two or more dice are available of different colors, we can handle tests of 36, 216, and so on, runs. If the red die is units and the green is tens, then green 3 and red 1 is a number 31. Runs can be given numbers 11 through 16, 21 through 26, and so on, and the plan developed through successive throws of the dice. Alternatively, and perhaps with greater dignity, a set of random number tables can be obtained and used to assign test-point sequence.[*]

What we are discussing here is *experimental control,* which, with precision, forms the crucial pair of desired aims in all test efforts. Although control and precision are related, it is perfectly possible to have excellent precision and terrible control or vice versa. Experiments in the social sciences are often very precise in the sense that the numerical counts of occurrences are exactly correct.

[*]For example, see Brownlee, 1953, pp. 21–22.

Yet the control over the behavior of human beings in experimental situations is so difficult as to be impossible in many such tests. On the other hand, a test such as the detection of the rotation rate of Venus by spectrographic studies of its light is very imprecise yet entirely free (if the night is clear) of other effects.

6-4 RANDOMIZED BLOCKS: EXTRANEOUS VARIABLES

We have so far restricted our considerations in this chapter to those simplest of experiments having one independent or controlled variable X and a dependent result R. We shall classify this as a *one-factor experiment*. But although there may be only one controlled variable or factor, we would be naive if we did not recognize the possibility of other uncontrolled or *extraneous* variables. We have already suggested such variables in the form of changes in temperature, pressure, or humidity, or changes in the attitudes of data takers. Extraneous variables of this sort change in a continuous manner with time and are best controlled by the simple randomization of test points as described in Sec. 6-3. Other extraneous variables are discrete in character. Examples are groups of people, different machines or instruments, different "mill runs" or material batches, different days or weeks or seasons of the year,* and so on. All may have definite, though uncomputable, effects.

Since we cannot eliminate the effects of many extraneous variables or allow for their effects by computation, we try to minimize their effects through randomization, which will spread extraneous effects more or less equally over all runs. When discrete extraneous variables can be identified, we can use the concept of the *randomized block*, as will be developed in this section.

Suppose that we have a new cutting tool which we wish to test in a production situation. We wish to establish the optimum machine speed for this tool to maximize production rate while not exceeding some rejection percentage. This is then a one-factor experiment with machine speed X as the independent variable and the production-rate function R as the dependent result. But such an experiment has one obvious extraneous variable, the machine operator. If we have perhaps 20 machinists, how are we to select a typical or average machinist to run our test? Obviously we cannot do this. Machinists vary so much in skill, temperament, physical strength, and so on, that to select a single "average"

---

*In one case in the author's experience the month of August was a particularly poor time for running heat-transfer and air-flow tests in a temporary building located in a large open field. Thousands of potato bugs were swarming at this time, and they plugged orifice and pressure lines and blocked air passages in the heat-exchanger test units.

machinist to run the test would be foolish. Let us then, through some chance method, select four machinists, each of whom will run a given speed over an entire shift. To balance the test, let us select four different speeds, so that each machinist will run each of the four speeds over a 4-day period, and we can average the results of each speed run. This will randomize the test over the extraneous variable, *machinist.* Assigning the speeds numbers 1, 2, 3, and 4 and the mean letters *A, B, C,* and *D,* a possible plan might be

Machinist	Shift day			
	Monday	Tuesday	Wednesday	Thursday
A	1	2	3	4
B	1	2	3	4
C	1	2	3	4
D	1	2	3	4

Such a plan is surely defective, for it ignores the effect of sequence on the test. The enthusiasm, interest, or perhaps dismay generated by a new tool on Monday shift may pall by Thursday, and production may fall for this reason. Alternatively, learning may occur, and production may go up. We have not randomized the extraneous variable, shift day. Suppose that we draw speed numbers out of a hat for each machinist, thereby randomizing the sequence with which they occur:

Machinist	Shift day			
	Monday	Tuesday	Wednesday	Thursday
A	4	2	1	3
B	2	3	1	4
C	3	2	1	4
D	1	3	4	2

This is an improved plan, but we can do better. Notice that in this hit-or-miss randomizing method, speeds 1 and 4 fall mainly in the last two days. Thus a decrease in interest at the end of the test might suggest a peaking in the middle speed range that is not actually a result of speed changes at all. Let us completely randomize this test such that each speed appears only once in a given day and no machinist runs at the same speed more than one day. Such a plan might be

	Shift day			
Machinist	Monday	Tuesday	Wednesday	Thursday
A	1	2	3	4
B	3	4	1	2
C	2	1	4	3
D	4	3	2	1

where we have constructed a *Latin square,* a special type of experimental plan that is part of the general family of *factorial experiments,* although factorial experimentation implies not only random plans but also the analysis of the results using advanced statistical methods. In engineering, we may or may not be interested in the actual effect of extraneous variables on our desired result. Here, for example, we could through statistical means investigate the relative strengths of shift day and machinist on our production rate and draw certain conclusions about the interaction of these variables, their variance, and so on. This is discussed in Sec. 9-5.

We may be able to improve our machine tool experiment still further. To let each machinist stay with a given machine—and machines may show large differences among themselves—might introduce bias due to such machine differences. Taking four machines as *W, X, Y,* and *Z,* we wish to distribute the test runs among the machines such that each machine is used only once by each machinist and is used only once with each speed. Then

	Shift day			
Machinist	Monday	Tuesday	Wednesday	Thursday
A	$1W$	$2X$	$3Z$	$4Y$
B	$3X$	$4W$	$1Y$	$2Z$
C	$2Y$	$1Z$	$4X$	$3W$
D	$4Z$	$3Y$	$2W$	$1X$

which will permit us to average out the effects of shift day, machine, and machinist using a *Graeco-Latin square.* There could be yet another extraneous variable, such as steel lot, but the addition of this variable to our plan is left as an exercise. Such "high-order" squares have been little used in experimentation, probably because of the great difficulty of getting five or more variables in desired combinations and all at the same number of levels. Interestingly enough, a 6×6 square is possible only with three variables, that is, as a Latin square.

When the four days and the 16 runs are complete, we would very likely be satisfied with a numerical average of the four production figures at each speed, plotted against speed. In Chap. 9 we shall see how a *test for significance* can be made on such data for cases when the effect of speed is not obvious from the usual engineering curve.

The square array is not necessarily the most convenient form for many experiments. We might, for example, wish to vary the speed over six increments, yet use fewer machinists and machines. A great many partial and unbalanced experimental designs are available in the literature, under the names *Youden squares, lattice squares*, and so on.[*] For common engineering use, multiple Graeco-Latin plans are often sufficient and are easily applied. For example, with speeds 1, 2, 3, 4, 5, and 6; machinists *A, B, C*; and machines *X, Y,* and *Z,* we can construct two 3 X 3 squares:

	Shift day			Shift day		
Machinist	Monday	Tuesday	Wednesday	Thursday	Friday	Monday
A	1X	3Z	5Y	2X	4Z	6Y
B	3Y	5X	1Z	4Y	6X	2Z
C	5Z	1Y	3X	6Z	2Y	4X

where we have distributed six speeds equally to the first and second blocks, "interlacing" as far as possible. This plan is not so randomized as a single 36-item square including six machinists, six machines, and six days, but would probably suffice.

> **Example 6-2** We wish to subject a series of transistor radios to operation at six different temperatures and plot loss of sensitivity at 100 h against temperature level. Radios from two different plants, with printed circuits of three kinds of plastic, are to be tested. The humidity is extraneous and cannot be controlled, and only two test chambers are available. What is a good plan?
>
> *Solution* A short test will result if we use two 3 X 3 squares, with each square taking radios from a different plant. Let temperature levels be T_1 through T_6 and types of plastics be P_1 through P_3, and let us set up 100-h periods *A, B,* and *C.* Then

[*]Kitagawa and Mitome, 1953.

Type of plastic	Plant 1			Plant 2		
	T_1	T_3	T_5	T_2	T_4	T_6
P_1	A	B	C	A	B	C
P_2	B	C	A	B	C	A
P_3	C	A	B	C	A	B

might be a plan. But, although this is about minimum in size for a factorial experiment in this case, the time is long to completion. We can place only a single radio in each of the two furnaces for each 100-h period, and the entire test will take 900 h. We might be satisfied simply to distribute the time periods in a less random manner in the interests of rapid completion. The design then is not a pair of Latin squares:

Type of plastic	Plant 1			Plant 2		
	T_1	T_3	T_5	T_2	T_4	T_6
P_1	A	C	B	B	A	C
P_2	A	C	B	B	A	C
P_3	A	C	B	B	A	C

Now three radios can be placed in each furnace, and the test totals only 300 h in length. We pay for this short time in poorer control over a humidity effect, for now six radios undergo testing simultaneously. By interlacing, that is, putting temperatures T_1 and T_4, T_3 and T_6, and T_5 and T_2 together, we partially mitigate any effect of humidity on the temperature–versus–sensitivity-loss curve.

6-5 MULTIFACTOR EXPERIMENTS: CLASSICAL PLANS

Many tests involve two or more controlled and variable factors, and we shall refer to them as two-factor, three-factor, and so on, experiments. In all such experiments, one, two, or many extraneous variables may also be present. In such multifactor tests, we often have the choice between two types of experimental plans, *classical* or *factorial.* The classical plan is in almost universal use by engineers everywhere and is perfectly general in application. The factorial plan is often shorter, always more accurate (for a given length of test), but has much less general application.

If we are given a dependent result R that is a function of several independent variables, X, Y, Z, and so on, the basic classical plan consists of holding all but one of the independent variables constant and changing this one variable over its range, following any spacing plan that we have worked out and allowing for extraneous variables. If the mathematical relationship among the independent variables is simple, this should reveal the function of R versus the changing variable (say, X). Then all but the next variable (say, Y) may be held constant, and Y may be varied to find the separate RY function. Essentially, a multifactor classical experiment is simply a series of one-factor experiments. This limited classical approach will find such simple functions as

$$R = AY^n + BX^m$$

$$R = AY^n X^m$$

$$R = AYB^{cX}$$

and so on. For a two-factor experiment with each factor taken over five levels, we can diagram the plan as follows:

Y level

		1	2	3	4	5
	5			*		
	4			*		
X level	3	*	*	*	*	*
	2			*		
	1			*		

where the asterisks indicate configurations of the test apparatus that are to be run. When the function is more complicated, such as

$$R = AX \sin \frac{BY}{X}$$

$$R = A + BX^m Y^n + CX^o Y^o$$

$$R = AX^{bY}$$

it is doubtful that a limited plan of X and Y at one level each would unravel it, and we might have to try X and Y at several levels; for example,

$$Y \text{ level}$$

	1	2	3	4	5
5	*	*	*	*	*
4	*		*		*
X level 3	*	*	*	*	*
2	*		*		*
1	*	*	*	*	*

Or we might completely fill this plan and run all 25 test points. The actual establishment of functions from data will be taken up in Chap. 7. When a classical experiment, either partial or complete, is planned, it does not have to be balanced. That is, we can choose ten X levels and only three Y levels if it is felt that the R-versus-X function is the more important or more difficult. In the test of heat exchangers, for example, correlation is often achieved by

$$N_{st} = kN_{re}^{a}N_{pr}^{b}$$

where the Stanton number N_{st} is the dependent result, and the Reynolds number N_{re} and Prandtl number N_{pr} are the two independent variables. In most practical situations, the Prandtl number changes very slightly over a wide range of temperatures, whereas the Reynolds number, which contains the fluid velocity, will show wide variation. In such a case we would vary the Prandtl number over far fewer levels than the Reynolds number. In the practical use of the final experimental equation, accuracy in the Stanton-versus-Reynolds number function is by far the most critical.

6-6 MULTIFACTOR EXPERIMENTS: FACTORIAL PLANS

We have already seen how Graeco-Latin square factorial plans can be applied to the one-factor test with several extraneous variables. It is also possible to apply these plans to engineering experiments of several factors, provided that certain limitations and precautions are observed. The special advantages of such factorial experiments over the classical types will become evident as we examine the methods.

The most serious restriction on the use of nonstatistical factorial experiments in engineering work is that only two types of general experimental functions can be readily handled. Furthermore, we must know which class of function we are dealing with before the data are processed. The first class is that in which the dependent result R is a function of the sums of the functions of the independent variables. This case has the general formula

$$R = f_1(X) + f_2(Y) + f_3(Z) \tag{6-2}$$

where f_1, f_2, and f_3 are functions of any level of complexity. This class of noninteracting relationship is very uncommon in engineering and physical science. In agriculture, such a relation is often assumed in problems involving such variables as depth of planting, amount of fertilizer, and seed concentration.

The much more usual second class of general relationship that can be handled by a factorial experiment is the case of the result being a function of the product of the individual functions of the independent variables, or

$$R = f_1(X)f_2(Y)f_3(Z) \tag{6-3}$$

This can be treated as a special case of the first class, because Eq. 6-3 transforms to the form of Eq. 6-2 if we take logs,

$$\log R = \log f_1(X) + \log f_2(Y) + \log f_3(Z) \tag{6-4}$$

Equation 6-3 is one of the most important general relations in scientific work. It includes the commonly assumed result in dimensional analysis,

$$R = kX^a Y^b Z^c \tag{6-5}$$

as well as a variety of complex forms such as

$$R = kX^a y^b e^{cZ}$$

or

$$R = \frac{k}{X} A^y \sin BZ$$

Examples of functions that are *not* in this class are

$$R = AX^a + Y^b Z^c$$

or

$$R = AX^a e^{bY/Z}$$

and an infinity of other functions having higher order complexity.

Let us now see how a factorial experiment could be run when the function is known to be of the class defined by Eq. 6-3. We shall consider a balanced experiment involving X, Y, and Z at three levels such that the Latin square is

	Y_1	Y_2	Y_3
X_3	Z_1	Z_2	Z_3
X_2	Z_2	Z_3	Z_1
X_1	Z_3	Z_1	Z_2

Let us assume that we know (from theory, intuition, or past experience) that Eq. 6-3 is the general expression governing the effect of X, Y, and Z on R. Let us write the three equations covering the horizontal X_1 row, but in logarithmic or noninteracting form,

$$(\log R)_a = \log f_1(X_1) + \log f_2(Y_1) + \log f_3(Z_3) \tag{6-6a}$$

$$(\log R)_b = \log f_1(X_1) + \log f_2(Y_2) + \log f_3(Z_1) \tag{6-6b}$$

$$(\log R)_c = \log f_1(X_1) + \log f_2(Y_3) + \log f_3(Z_2) \tag{6-6c}$$

Now let us add these three equations together, obtaining

$$\Sigma \log R_{x1} = 3 \log f_1(X_1) + \log (f_2(Y_1)f_2(Y_2)f_2(Y_3))$$
$$+ \log (f_3(Z_3)f_3(Z_2)f_3(Z_1))$$

We can repeat the same procedure for the middle, or X_2 row, obtaining

$$\Sigma \log R_{x2} = 3 \log f_1(X_2) + \log (f_2(Y_1)f_2(Y_2)f_2(Y_3))$$
$$+ \log (f_3(Z_3)f_3(Z_2)f_3(Z_1))$$

and similarly for the top, or X_3, row. The above equations can be written

$$\log f_1(X_1) = \frac{\Sigma \log R_{x1}}{n} - \text{const} \qquad (6\text{-}7a)$$

and

$$\log f_1(X_2) = \frac{\Sigma \log R_{x2}}{n} - \text{const} \qquad (6\text{-}7b)$$

and so on for the X_3 level; n for a 3×3 square is 3, and for a higher level square is equal to the number of levels. What we have done in this proof is to show that, if the logarithms of the results are numerically averaged over a single X, Y, or Z level, the effects of those factors that are changing (Y and Z in the case examined) will remain the same from one X level to the next. Thus, all changes in the log average of the result are due wholly to the effect of X alone. We could easily continue to show the same result when averaging occurs over the three Y levels and then over the three Z levels. If yet another variable, say, W, were added, forming a Graeco-Latin square, the same rule would apply in finding the effect of W on R.

If the experimental function is known before the experiment is analyzed to be a sum type following Eq. 6-2, we obtain the effect of X, Y, and Z on R by *averaging the appropriate R values rather than log R.* If it is not known which class applies or whether either applies at all, it is recommended that this factorial approach not be used and that a standard classical approach be taken.

The analysis of the various functions can be undertaken using plots of log X versus log R_{avg} or by taking antilogarithms and examining the X-versus-R_{avg} function or by numerical means.

Suppose that we obtain tabulations or curves of R as a function of X, Y, and Z separately. Equations 6-7a and 6-7b show that such curves or tabulations will yield us

$$R_x = k f_1(X)$$

$$R_y = k' f_2(Y)$$

$$R_x = k'' f_3(Z)$$

where R_x is the antilog of $\Sigma \log R_x/n$, k is the constant in Eqs. 6-7a, 6-7b, and so on, made up of the Y and Z portions eliminated through use of the Latin square, and $f_1(X)$ is the function of the X variable, as noted. If we solve these three equations for $f_1(X)$, $f_2(Y)$, and $f_3(Z)$ and substitute in Eq. 6-3, we obtain

$$R = K(R_x)(R_y)(R_z) \tag{6-8}$$

where K is $(kk'k'')^{-1}$. We can evaluate K if we know the final result R and the individual Rs from the X, Y, and Z curves or tabulations. The following example will show the method.

Example 6-3 A student group wished to study the effects of speed, load, and cooling-water temperature on the operating characteristics of a Dodge internal-combustion engine mounted on a test stand. From their study of internal-combustion engine testing, they assumed that such characteristics are related by a products-of-functions type of experimental equation, as in Eq. 6-3, so that a Latin square is possible with the results log-averaged. How was this test planned, and what sort of data resulted?

Solution The students selected a 4 X 4 square having the construction

Temperature in square (°F)

Speed (rpm)	1400	1600	1800	2000
Dynamometer load (lb)				
87.5	110	135	160	200
66.0	200	110	135	160
44.0	160	200	110	135
22.0	135	160	200	110

after first ensuring that all 16 of these engine configurations could be met by the test equipment. Note that all Latin squares are not necessarily possible. A square with the top row having temperatures 200, 160, 135, and 110°F reading from left to right was not possible, since the cooling-water flow was insufficient to hold temperature at its minimum (110°F) when load and speed were at maximum.

The 16 runs were then made, and the square was filled with the basic dependent result, which was fuel consumption in pounds of fuel per hour:

21.2	24.5	28	29
14	16	19	22
8	11	14	16
6	8	8	12

Now if the general equation (6-3) governs this test, we must make a logarithmic average and then take the antilog (of fuel consumption) as follows:

	Log (fuel consumption)				Load varying		
					Sum	Avg	Antilog
$T = 110°F$	1.326	1.392	1.447	1.462	5.625	1.406	25.5
$T = 200°F$	1.146	1.204	1.279	1.342	4.971	1.243	17.5
$T = 160°F$	0.903	1.041	1.146	1.204	4.295	1.072	11.9
$T = 135°F$	0.778	0.903	0.903	1.079	3.664	.916	8.2
Sum	4.154	4.538	4.775	5.088	4.756	1.189	15.5 ($T = 110°F$)
Avg	1.038	1.135	1.194	1.272	4.650	1.163	14.5 ($T = 135°F$)
Antilog	10.9	13.6	15.6	18.7	4.596	1.149	14.1 ($T = 160°F$)
		Speed varying			4.553	1.138	13.8 ($T = 200°F$)
					Temp varying		

Fuel consumption, specific fuel consumption, and efficiency were then found, and Figs. 6-5, 6-6, and 6-7 were plotted to show the results of the test. We cannot use these curves directly to give us, say, the efficiency at a given load, for they represent averages rather than discrete values. Let us take Eq. 6-8 and compute the unknown constant K_1 and then use this equation to interpret the curves. The data in the upper row, second column from the left in the Latin square, have the values: load, 87.5 lb; rpm, 1600; temperature, 135°F; and fuel consumption for this run, 24.5 lb/h. Now we see, by looking at either the fuel-consumption curves or the analyzed square

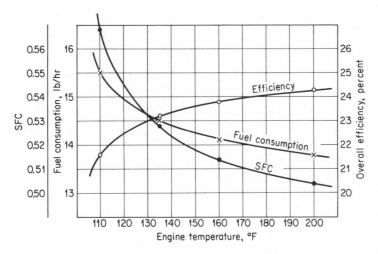

Figure 6-5 Plot showing the effect of engine temperature on fuel consumption and other operating parameters of the internal-combustion engine test discussed in Example 6-3.

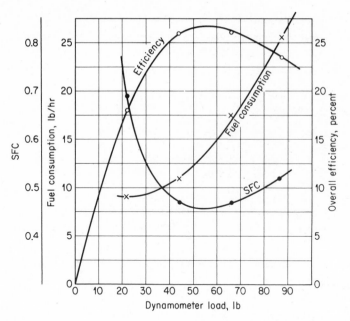

Figure 6-6 Plot showing the effect of dynamometer load variation on fuel consumption and other operating parameters of the test noted in Example 6-3.

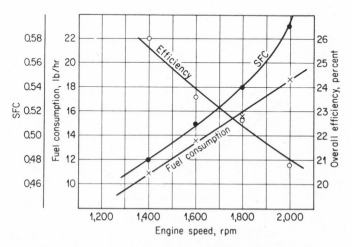

Figure 6-7 Plot showing the effect of engine speed on fuel consumption and other operating parameters of the test noted in Example 6-3.

itself, that a load of 87.5 lb gives an average fuel consumption of 24.5 and that this fuel consumption is the direct result of the log-averaging process, which we showed would eliminate the effect of rpm and temperature variations. Similarly, the average fuel consumption for 1600 rpm is 13.6 and for 135°F is 14.5. Thus from Eq. 6-8 we find

$$K = \frac{24.5}{25.5 \times 13.6 \times 14.5} = 0.00488$$

But notice that we can check this K value by repeating this calculation on any other point of the Latin square. For example, the lower right-hand run (2000 rpm, 110°F, 22.0 lb) gives a K of $12/(8.2 \times 18.7 \times 15.5)$ or 0.0051. The following square shows the K values computed for each of the 16 runs:

0.0049	0.0049	0.0050	0.0044
0.0053	0.0043	0.0048	0.0048
0.0044	0.0046	0.0049	0.0050
0.0048	0.0051	0.0052	0.0051

The differences in the various Ks are indications of how badly the data deviate from the ideal of Eq. 6-3. These deviations may be a result of the failure of Eq. 6-3 to predict exactly the functional relationships of the test variables or the lack of control in holding the variables at their planned levels, or simple lack of precision in the measurements. The average K for these 16 runs is 0.00485, and the maximum deviation from this is 0.00055, or 11%. In this test, most of this deviation is probably due to the difficulty in control of speed and temperature while fuel-consumption readings are being taken. Using this average K, we could now answer a number of questions about the apparatus. For example, we note that maximum fuel consumption occurs at maximum speed (2000 rpm), minimum temperature (110°F), and maximum load (87.5 lb). Then

$$(FC)_{max} = 0.00485 \times 25.5 \times 15.5 \times 18.7 = 35.6 \text{ lb/h}$$

which is the fuel consumption at 2000 rpm, 110°F, and 87.5 lb, a condition that was not actually run. The uncertainty is about ±11%, with this uncertainty figure including perhaps 90 to 95% of all the data.

A similar type of computation can be made for specific fuel consumption or efficiency. The material in this example is used in Example 7-3 to show how further analysis is possible using graphical techniques.

It might be asked just what has been gained by this rather involved type of plan. In a balanced, three-level experiment involving three variables, we can obtain three three-point curves with only seven runs, instead of the nine required for the Latin square. These seven can be represented in three-dimensional space as

where the three axes have a common point or run. Even when a 3×3 Graeco-Latin design is used, a classical test of nine runs will give the same number of curves and points and handle the same number of variables. But notice that each three-point curve obtained from a factorial test involves all nine runs. Each curve point is actually an average of three separate determinations. From our discussion of Eq. 2-25 on the improvement in precision of a mean by taking increased numbers of readings, we would expect that each factorial test point would be $3^{1/2}$ or 1.7 times as precise as each classical test point. In a 4×4 experiment, we would have to replicate each run four times in a classical experiment to achieve a precision equal to a 16-run factorial plan. If three variables are involved, the basic classical plan would involve 10 runs,

which when completely replicated four times gives 40. Thus, the main advantage of the factorial multifactor experiment is increased precision with little or no increase in the amount of testing. The main disadvantages are the problems associated with first selecting and then running a large number of test runs with perhaps imperfect knowledge of the operating envelope of the equipment, plus the fact noted in Example 6-3 that the final curves are not completely useful alone but must be changed to general functional relationships.

The approach to factorial experimentation taken here is essentially non-statistical. In the case of optimizing a machine tool, we did not care what relationship existed between output and shift day or machinist but only wanted to randomize these effects out of the test. In the engine test, we knew at the beginning that strong functional relationships existed between the dependent result and the three variable parameters of the test. In many situations, we wish to decide if an effect is present but suspect that such an effect is not as strong as is suggested in Figs. 6-5 through 6-7. When this is the case, factorial tests are often used, but the analysis of the result squares are statistical in nature. Entire books are available on this type of experimental plan. The last section in Chap. 9 will introduce this kind of statistical analysis.

Without question, the bulk of engineering experimentation will continue to be performed along classical lines. In cases where the function is known to be a product type, where several variables are involved, and where the precision is thought to be poor, the engineer should always consider the factorial approach. And, as we shall see in Sec. 9-5, the factorial plan is essential for multifactor experiments of low precision.

6-7 SUMMARY

In this chapter we have considered a number of pretest planning methods. Starting with a series of known, independent variables, we first should consider the spacing of test runs or points over the envelope of the apparatus. The basic concern in such spacing is to ensure that the final plotted curve is equally precise over its range. When an uncertainty analysis suggests that certain portions of a curve or map have less precision than other portions, we try to adjust the point spacing so that more points will fall in the bad-precision area. Ideally, we would like to have (approximately) the lack of precision squared in these areas so that if the poorest area is one-half as precise as the best area, we would put four times as many test points in the less precise area. Often, however, it is not possible to reach this ideal.

When the spacing of runs is not dictated by regions of varying precision, we wish to prevent the wasteful "bunching" of test points by setting up equal data spacing along the curve(s). In many cases, where the form of the function is partially known, the investigator can algebraically transform the equation and create a linear form, making the job of spacing quite easy. This approach means, however, that the equation should be plotted in its transformed form, something that is not always desired.

After the actual levels of the independent variables are chosen, the investigator then has the choice of a sequential plan, in which the independent variable changes proceed in a regular increasing or decreasing sequence, or a random plan, in which the chosen variable levels are selected at random, following some irregular plan derived from dice throws, card drawings, or the use of random-number tables. The sequential plan is usually essential for those irreversible tests in which progressive changes or deterioration occurs and in certain special cases where the sequence of data taking is a parameter of the test in its own right. For the majority of reversible engineering experiments, the random plan will average out any regular variations due to surroundings, test personnel, and test apparatus defects and is thus very desirable.

When these extraneous variables of personality, weather, shift day, season, machine, material batch, and so on, can be placed in well-defined categories, their effects are best eliminated by using a factorial plan of the Graeco-Latin type, where a variety of extraneous variables are randomly distributed to the different runs. This can be done in balanced or unbalanced plans, although only balanced plans are considered in this book. Often a large block—six, eight, or 10 levels on a side—can be replaced by two or more smaller blocks with little loss in randomization.

When an experiment involving several variable factors is to be run, we have the choice of the classical plan, in which all but one of the variables are held constant at one level, or the factorial plan, in which all the variables are changed in each run as dictated by an appropriate Graeco-Latin square. The classical plan is in universal use by engineers, is general in application and not difficult, but may be lengthy if an experiment has poor precision and must have its data replicated several times. The factorial plan is restricted here to cases where the dependent result is equal to the sums of functions of the variables or to the products of functions of the variables, and we must know which. The product-of-functions case is very common in engineering, and if we are sure that an experimental function has this form, we need only make numerical averages of the logs of the results to reveal the effect of each variable alone on the results. This is all provided, of course, that we have operated the apparatus to follow a randomized plan. The main advantage of using this factorial approach is that the entire set of data is used for each curve, and the precision is therefore at a maximum.

We have barely scratched the surface of test planning. The medical, biological, and chemical sciences have pioneered sophisticated and complex experimental designs. Much of this material is, for one reason or another, not practical for mechanical and electrical experimenters in its present form. Still, engineers have lagged behind their colleagues in biology and agriculture, and many radical improvements in engineering test planning can be expected.

PROBLEMS

6-1 In the test of a steam-to-water heat exchanger, the relationship among the temperatures, the water flow rate w, and the overall heat-transfer coefficient U is given by

$$\frac{\Delta T_{water}}{T_{steam} - T_{water\ in}} = 1 - e^{-UA/wCp}$$

where A is 10 ft^2 and C_p is 1 Btu/lb $\cdot$ °F. The apparatus will permit a low-water flow of 20 lb/h, which gives a water-out temperature of 130°F. High flow is 500 lb/h with a water-out temperature of 54°F. Steam temperature and water-in temperature are fixed at 200 and 50°F. Select four intermediate values of w so that log U versus log w will show even spacing, assuming that this plot is a straight line.

6-2 The general equation for ship speed versus engine horsepower is $hp = a + bV^3$, where a and b are constants relating to ship form and propeller type. At 5 knots the ship requires 29 hp at the shaft, whereas at 12 knots, 230 hp is needed. Locate three or four intermediate speeds such that the intervals are equally spaced on a plot of log (speed) versus log (hp).

6-3 A microwave focusing antenna takes power from a transmitter at distance X and focuses this power on a stationary platform at altitude D. The usual optical equation $D = X/(a + bX)$ applies. When X varies from 21 to 29 m, D varies from 32,000 to 6000 m. Decide how this test should be run to give a linear plot and place six points between these extreme values to give even spacing on the linear plot.

6-4 A hydraulic transmission is to be tested by applying a known horsepower to one end of the line of pipes and then measuring the useful power, uhp at the other end. The expected function is $uhp = hp - 1 \times 10^{-5}(hp)^3$, and we wish to vary hp between 100 and 200. Decide how to vary hp over five steps so that a curve of uhp versus hp suitable for equipment buyers is prepared.

6-5 Devise suitable experimental plans for the following tests:

(*a*) Four types of specimens are to be torsion tested to find yield point and total angular displacement at fracture. We wish to examine the effect of two different controlled loading speeds on these variables and attempt to minimize the extraneous effects of four different people putting the specimen in the jaws and the four different batches of aluminum that are used.

(*b*) Two internal combustion fuel-test engines are to be used to check three fuel samples for octane rating. One engine tests one shift, and we have six different technicians available. We wish to randomize completely the effect of technician, shift, and engine. What is the minimum number of shifts needed to accomplish this? Draw a test plan.

(*c*) An airfoil cascade is to be tested of six stagger angles, six incidence angles, and three solidity ratios. Design an experiment to find the effect of these variables on flutter velocity if (1) The function is simple but not a sum or power type; (2) The function has very complex interaction; (3) The function is of the product-of-functions type and has very poor precision.

6-6 Draw a 5 × 5 Graeco-Latin square that is *completely* different from the one given in Appendix C (do not simply change the position of rows or columns).

6-7 In the example involving a 4 × 4 Graeco-Latin square considered in Sec. 6-4, we wish to randomize along with shift day, machine, and machinist the extraneous variable steel lot, the four lots having designations P, Q, R, and S. Draw the new design. Can yet another variable be randomized along the 16 runs of this test?

6-8 Starting with the 5 × 5 Graeco-Latin square in Appendix C, add new sets of variables F, G, H, I, and J, K, L, M, N, and O, and so on, until no further sets can be added in a randomized manner. How many different extraneous variables could we randomize with a 5 × 5 square in a one-factor experiment?

6-9 Devise suitable experimental plans for the following tests:

(*a*) The melting point of a variable mixture of five components A, B, C, D, and E, follows the function $MP = (a\% A + b\% B + c\% C + d\% D + e\% E)k$, where a, b, c, d, e, and k are constants to be found. Design an experiment with each component at four different levels plus 0%.

(*b*) Suppose that, in the test of rotating equipment, we have as extraneous variables two different shifts and four methods of taking eight levels of rotating speed. Design a test with no more than 32 runs.

(*c*) Suppose that in the test in (*b*) we wish to obtain additional runs at the lowest speed where precision is poor, but we wish to randomize data taking and shift and include the effect of temperature. How many runs are needed for this substitute plan, and what does it look like?

6-10 Prove that when a result R is equal to the sums of individual functions of the variables, we can numerically average the result at each level to find the separate functions of R and the variables.

6-11 Can the following physical equations be investigated by a factorial experiment? Show algebraic proof of your answer, and if it is "yes," explain how the columns should be averaged. Draw a 3 × 3 square for those equations that are possible in a factorial test.

(*a*) $C = Ak/\ln(d/r)$, where C is capacity, k is the dielectric constant, and d and r are geometric variables with A constant throughout.

(*b*) $I = Aw(a^n + b^m)$, where I is the moment of inertia, w is the weight, a and b are body dimensions, and A, n, and m are constant.

(*c*)

$$P = \frac{Ast}{R + t}$$

where

$$P = \text{allowable pressure}$$

$$s = \text{material working stress}$$

$$R = \text{inner radius}$$

$$t = \text{wall thickness}$$

$$A = \text{a constant}$$

6-12 A machine is operated on a factorial plan with three independent variables: flow w, rotational speed rpm, and back pressure P. The final result of each run is efficiency E. The following points are taken:

rpm	w	P	E
1000	14.2	26	12.6
1260	14.2	158	25
1590	14.2	63	12.5
1260	17.8	63	24.8
1590	17.8	26	12.8
1000	22.3	63	51
1000	17.8	158	50
1590	22.3	158	49.5
1260	22.3	26	25.5

Arrange these data in a Latin square and decide whether a log average or a plain average will give the most logical results. By proper averaging, find the function of E with respect to the three variables. (*Hint*: The functions should all plot straight lines on log–log paper.)

6-13 Two ovens are available to test four different kinds of transistors having two different brands of lead wires. A single test requires a complete day, but the ovens are identical and interchangeable (not extraneous variables). We wish to randomize days of the week, kinds of transistors, oven temperature (four values are ample), and wire type. Design a test plan to do this, and state the maximum number of transistor specimens needed and the minimum number of days.

6-14 A series of automatic, electronic flue-gas analyzers that sense percent oxygen and percent nitrogen and read out the percent excess air E in the combustion chamber are to be tested over their operating range. This test is to be conducted with a series of premixed gas samples containing various amounts of O_2 from 0 to 15% and a single N_2 percentage of 80% with CO_2 making up the remainder. If the governing equation is

$$E = \frac{O_2}{0.264 N_2 - O_2}$$

with the symbols O_2, N_2, and E standing for decimal fractions, and we wish to have a dozen samples to cover the applicable range, what spacing of O_2 should we choose?

6-15 A series of electric heating elements having voltage V and current I is tested as to rapidity of heating a liquid metal of C_p 0.19 Btu/lb $\cdot$ °F, mass M, and temperature rise in 1 h, ΔT. The entire system is carefully insulated, and these data result:

V (volts)	10	10	10
I (amperes)	1	3	4
M (lb)	2	4	4
T (°F)	95	140	179

Decide whether further tests should involve larger currents and temperatures or smaller values of these quantities. What is the percent error of each balance?

BIBLIOGRAPHY

Brownlee, K. A.: *Industrial Experimentation,* Chemical, New York, 1953.

Chew, V.: *Experimental Designs in Industry,* Wiley, New York, 1958.

Cochran, W. G., and G. M. Cox: *Experimental Designs,* Wiley, New York, 1950.

Fisher, R. A.: *The Design of Experiments,* Oliver & Boyd, London, 1949.

Hicks, C. R.: *Fundamental Concepts in the Design Experiments,* Holt, Rinehart and Winston, New York, 1964.

Kitagawa, T., and M. Mitome: *Tables for the Design of Factorial Experiments,* Dover, New York, 1953.

Mann, H. B.: *Analysis and Design of Experiments,* Dover, New York, 1949.

Pratt and Whitney Aircraft, Engineering Statistical Methods Group: *Increasing and Efficiency of Development Tesint, PWA-2236,* East Hartford, Conn., July, 1963.

Wilson, E. B.: *An Introduction to Scientific Research,* chap. 4, McGraw-Hill, New York, 1952.

GRAPHICAL AND MATHEMATICAL
ANALYSIS: DATA OF HIGH PRECISION

By now it should be evident that graphs and graphical analysis are at the core of the study of most engineering test data. This is not to say that considerable mathematical effort may not be expended on the analysis of a test (as suggested in Sec. 4-6), but that all such analysis starts with forming appropriate graphs. It then often proceeds with the mathematical analysis of such graphs, using the methods of *rectification, function discovery,* and *extrapolation,* which we shall discuss in this chapter. In addition, we shall consider ways in which sets of high-precision data can be *graphically compared.*

In this chapter we shall deal with data that has *high precision,* which we define as follows: *The data must be sufficiently precise to define the experimental function, its mathematical character, and its extensions.* One example of such graphs already discussed are those in Fig. 2-2. Although each of the gauges calibrated there had some precision error, mainly because of their hysteresis characteristics, we could still readily determine that the bourdon gauge has a relatively constant offset characteristic or sum-type error, and the capillary gauge has an error proportional to the reading or product-type error. Figures 4-7, 6-3, and 6-5 through 6-7 are all examples of data of high precision that can (and some of which will) be analyzed by the methods of this chapter.

Chapter 8 will deal with graphical plus statistical analysis of data that has *low precision,* that is, to which we must apply some statistical methods to prove that a functional relationship exists, although we shall usually not be able to define this relationship exactly. The data in Chap. 2 involving fuses is an example of data that requires low-precision methods. If, for example, we were to test the fuses in Sec. 2-3 at several different loading rates (Prob. 2-10 shows one example of the data from such a changed-rate test), it is doubtful that we could obtain the exact functional relationship between loading rate and blow-out current. We would probably have to apply a statistical curve fit, such as the least-squares method described in Chap. 8, to these data and fit a straight line to it as best we could.

In Chap. 9, we shall consider methods of data comparison and analysis that involve "pure" numbers, that is, statistical samples, although even there, some insight can be gained by graphical presentations.

7-1 GRAPHICAL RECTIFICATION OF TEST DATA

We shall assume in these next two chapters that the plan of our experiment is such that its results can be represented as a series of *dependent variable results,* usually plotted on the Y axis (*ordinate*) against the *independent variable variations,* usually plotted on the X axis (*abscissa*). Even if the test involves more than one independent parameter, as is so often the case in the experiments in Chap. 4 based on dimensionless groups, we shall assume that all but one parameter is held constant, the *classical plan* of Sec. 6-5, or averaged over a *factorial* plan to allow an X-versus-Y presentation as in Sec. 6-6.

If such data form a smooth curve on whatever kind of coordinates are first tried, it is reasonable to assume that they have sufficiently high precision to permit *rectification,* followed by *function discovery.*

Rectification simply means using whatever mathematical transformation of either the two data sets or the X and Y axes of the graph that causes the data to plot a straight line. There is no more important experimental technique in this entire book!

Figure 2-7 is an example of rectification. By arranging the intervals on the Y axis in a particular way, and then plotting randomly deviating data in a cumulative manner, we achieve a straight line if the sample is normally distributed. And even when we do not achieve this line, as in Sec. 2-10, we may be as strongly led to discoveries as when the rectification succeeds.

The first step in analyzing a set of X-Y data, then, is to determine if it is regular by plotting on any convenient set of scales, usually on *linear* graph paper. If the data do not form a straight line on linear graph paper, the engineer may next try *log-log coordinates* (or taking the log of both the X and Y values and plotting on linear paper). Log-log plotting will straighten out or rectify the simple yet important function

$$Y = kX^a \qquad (7\text{-}1)$$

where the transformation is (with k and a as fitting constants)

$$\log Y = \log k + a \log X \qquad (7\text{-}2)$$

Next to the straight line, this is probably the most common form of experimental function and is often assumed in dimensional analysis where the pi groups, raised to powers, multiply.

A third common type of graph paper is the *semilogarithmic type*, with one scale in log form and the other linear. This will give a straight line if the data are following a law of the type

$$Y = k10^{ax}$$

which is identical with

$$Y = ke^{2.3026aX}$$

The transformation of these functions is

$$\log Y = \log k + aX \qquad (7\text{-}3)$$

and to achieve a straight line the Y axis must be logarithmic with the X axis linear.

A number of special forms of paper (i.e., trilinear, hyperbolic) can sometimes be found, but are not really needed. For example, the hyperbolic curve

$$Y = \frac{X}{a + bX} \qquad (7\text{-}4)$$

can be rectified by plotting $1/Y$ versus $1/X$ on linear coordinates.

Some types of data will give peaks or hollows on linear coordinates. This immediately suggests a parabolic function or more general *polynomial*

$$Y = a + bX + cX^2 \qquad (7\text{-}5)$$

which may have additional terms if there is more than one peak. For the three-term parabolic equation there are several means of rectification. Let (X_1, Y_1) be the coordinates of any point on a smooth curve through the data. Then

$$Y_1 = a + bX_1 + cX_1^2$$

which may be subtracted from Eq. 7-5 to give

$$Y - Y_1 = b(X - X_1) + c(X^2 - X_1^2)$$

We can divide both sides by $X - X_1$, obtaining

$$\frac{Y - Y_1}{X - X_1} = b + c(X + X_1)$$

$b + cX_1$ is constant, so that a plot of $(Y - Y_1)/(X - X_1)$ versus X will be a straight line if this is the equation governing the data.

Another approach is to differentiate Eq. 7-5,

$$\frac{dY}{dX} = b + 2cX$$

which suggests that a plot of $\Delta Y/\Delta X$ versus X will give a straight line if we take equal X intervals.

These and a number of other rectification rules are summarized in Table 7-1. The proofs of these rules are left as exercises.

It should be noted that one need not use only the actual data points in these tests. The best strategy is first to plot the data on linear coordinates and then to draw a smooth curve through the points. Then, select the function most

Table 7-1 Rectification rules

To test for the function:	Obtain a straight line when:
(a) $Y = aX + b$	X and Y are plotted on linear paper
(b) $Y = kX^a$	X and Y are plotted on log–log paper
(c) $Y = k(10)^{ax}$ or $Y = ke^{ax}$	X is plotted on the linear scale, Y on the log scale of semilog paper
(d) $Y = X/(a + bX)$ or $1/Y = a/X + b$	$1/Y$ and $1/X$, or X/Y and X are plotted on linear paper
(e) $Y = a + bX + cX^2$	$(Y - Y_1)/(X - X_1)$ and X are plotted on linear paper
(f) $Y = X/(a + bX) + c$	$(X - X_1)/(Y - Y_1)$ and X are plotted on linear paper
(g) $Y = k(10)^{bX + cX^2}$ or $Y = ke^{bX + cX^2}$	$(\log Y - \log Y_1)/(X - X_1)$ and X are plotted on linear paper

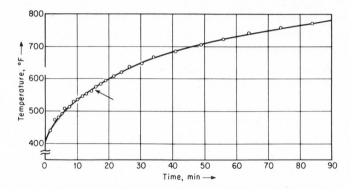

Figure 7-1 A plot of student data taken during an experiment on a heated stack. A typical "French curve" line drawn by eye through the points is shown, with an arrow indicating the time at which the potentiometer range change occurred. The reader may judge if one could have noted a discontinuity from such a plot.

likely to fit and use any X and Y on the curve needed to obtain a suitable confirmation of (or negation of) the hypothesized function. Notice that any such algebraic test should be carried out over the entire data range. Some experiments will give straight lines over a portion of the data envelope and curved lines over the remainder. Even when this occurs, however, attempted rectification of the data is essential. A curve having even a portion of its range plotting in a linear manner is far more revealing than one that curves over its entirety.

An actual example of this type of analysis is shown in Figs. 7-1 and 7-2. A

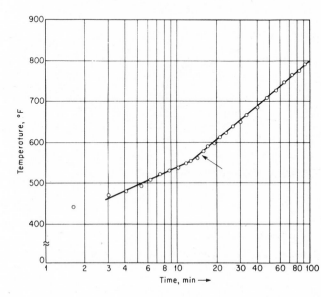

Figure 7-2 The data as shown in Fig. 7-1 but with the time axis logarithmic. The range-change point is again indicated by an arrow, and the abrupt slope change at this point is unmistakable.

senior mechanical-engineering experiment involved the taking of continuous readings of temperature at various points in a stack being heated. Figure 7-1 shows a conventional time–temperature plot from the thermocouple in the center of the stack, and it can be noted that a slight dip or flat region seems to exist between 500 and 600°F. Transient heat-flow data are often rectified by plotting the log of time versus temperature, as shown in Fig. 7-2. This semilog plot shows clearly the sharp change in the temperature-versus-log (time) curve slope at 550°F. Armed with such concrete data, it was a simple matter to check the original data sheet and note that at this temperature the potentiometer indicating the reading of the several couples was changed from low to high range. Thus, we can say with some assurance that the instrument is in error on at least one range and should be recalibrated.

Example 7-1 Figure 7-3 shows a simple experiment in which a subject attempts to exhale as strongly as possible through a tube of varying length L and constant diameter D, and the maximum static pressure observed on the inclined micromanometer is recorded. Since there is some variation because

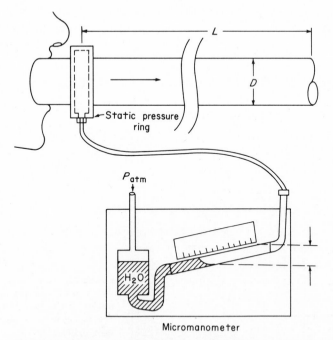

Micromanometer

Figure 7-3 Apparatus to obtain peak exhale pressure as a function of tube length. Note that unless the subject blows steadily for some time, this system will suffer response errors as noted in Chap. 5.

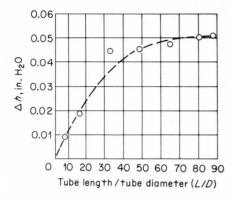

Figure 7-4 Linear-axis plot of experimental data in Fig. 7-3. Again, a "French curve" line is drawn ignoring the apparent "wild point."

of the difficulty of always exhaling the same way, data from 10 trials are averaged to obtain each L/D point. (See Prob. 2-37 for data on this experiment.) Figure 7-4 shows a *linear axis plot* of the results, with a "French curve" line drawn in. Can these data be rectified? Is the point at an L/D of 32.3 a "wild point," and does it invalidate our requirement of high-precision data?

Solution *Never question points until you have rectified the data.* Data involving fluid behavior, pressure drop, and velocity experiments are almost always best plotted on log–log coordinates. Figure 7-5 shows that the three data points for the short lengths form a straight line on these coordinates with a slope of almost unity, as we expect from a system in which pressure drop follows the usual *Fanning equation* (Eq. 4-1). Above an L/D of 33, the tube is so long that its volume contains a significant part of the breath, and the law expressing the maximum exhale pressure versus L/D has the new form shown.

Example 7-2 Rectification can be as useful in mechanics as it is in fluid mechanics. Figure 7-6 plots the plotted data from seven aluminum alloy columns, showing the maximum, that is, failure, load P, divided by the column area, versus the ratio of length to diameter. Theory suggests that columns fail in one of two modes, either by a crushing failure when they are short or by a bending instability when they are long. These long or Euler columns follows a law

$$\frac{P}{A} = \frac{K\pi^2 E}{(L/R)^2} \tag{7-6}$$

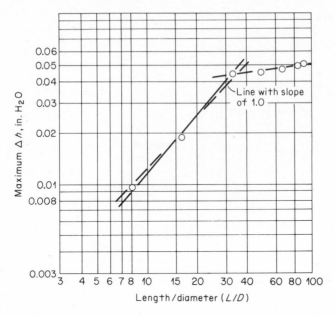

Figure 7-5 Log–log plot of data in Fig. 7-4, now suggesting two different data ranges.

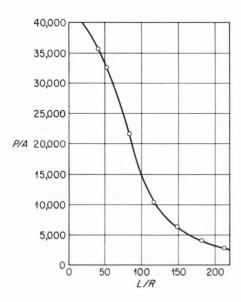

Figure 7-6 A linear-axis plot of column data with a curved line through the points.

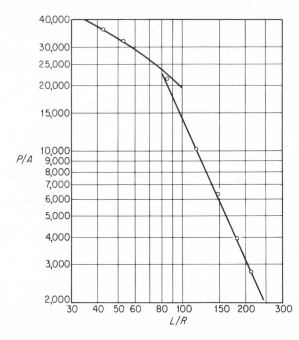

Figure 7-7 A log–log plot of the column data from Fig. 7-6, showing the straight-line or "Euler" region. The slope of these Euler points can be used to find other parameters of the test.

where K is an end-constraint constant and E is the modulus of elasticity of aluminum (10×10^6 psi). At what L/R value do these columns change their failure mode?

Solution The log–log plot, Fig. 7-7, shows the answer. The curve deviates from a straight line at an L/D of between 80 and 90. Beyond this value, the slope of the line is almost exactly -2.0, as Eq. 7-6 predicts. Thus the data rectification of Fig. 7-7 does two things (as did that of Fig. 7-5): It divides the data into zones or regions as anticipated by theoretical expectations, and it provides confirmation by *slope measurements* that theory and experiment are in fair agreement.

7-2 FUNCTION DISCOVERY

Once a partly or completely rectified graph has been found, the usual second step is to find the equation or *mathematical function* relating the independent and dependent variables. We began to do this in Examples 7-1 and 7-2 when we compared the *slopes* of these lines to the slopes predicted from theoretical expectations.

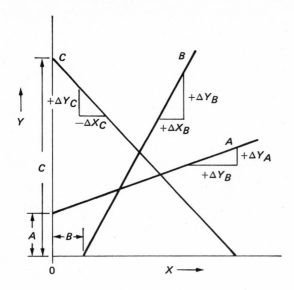

Figure 7-8 Graphical presentation of Eq. 7-7, the slope–intercept equation, showing the three possible variations.

Rectified data may pass through the origin of either the X or Y axis; that is, they may have an *intercept* on one of these axes, or as is the case when the data are rectified on log-log coordinates, there may be no intercept because neither scale goes to zero. When an intercept is present, we normally find the experimental function by applying the *slope–intercept equation*. Assuming that the X and Y data are all positive, there are three possible outcomes of our rectification, as shown in Fig. 7-8. (Note that vertical or horizontal lines indicate an *extraneous variable*; see Fig. 4-5.) The slope–intercept equations for these three cases are

$$Y_a = K_a x_a + a \tag{7-7a}$$

$$Y_b = K_b(x_b - b) \tag{7-7b}$$

$$Y_c = c - K_c x_c \tag{7-7c}$$

Example 7-3 In Example 6-2 a Latin-square experiment was outlined in which an auto engine was operated over a range of loads, speeds, and engine temperatures. One of the resulting curves was a plot of specific fuel consumption versus engine block temperature. Values from the smoothed curve are as follows:

Block temperature (°F)	120	130	140	150	160	170	180	190
Specific fuel consumption	0.545	0.532	0.523	0.518	0.513	0.51	0.507	0.505

What is the equation of this curve?

Solution Before attempting to fit these data, we should first wonder if we are dealing with the variables in their most significant engineering form. Experience with engines would lead us to expect somewhat better fuel consumption with increased temperature, for this is predicted by basic thermodynamic theory. But is the Fahrenheit temperature the best variable, or is the temperature level above the surrounding datum of more fundamental significance? Again we realize that when thermodynamics predicts better fuel consumption from a hot engine, this means hot in relation to the datum. In this test, the lab temperature was recorded as approximately 80°F; therefore let us fit SFC versus $T_{block} - 80$ or ΔT.

The function has no peaks or hollows, only a continuously decreasing SFC with increasing ΔT, with, however, a decreasing slope. This immediately suggests an exponential curve, (b) in Table 7-1, with the exponent a negative fraction. The check is to plot log SFC and $\log(T - 80)$ as shown in Fig. 7-9. This attempt fails to rectify the curve, and we must search for a more appropriate function. The hyperbolic curve (d) in Table 7-1 will also fit data of this sort. Let us test by obtaining 1/SFC and $1/\Delta T$,

1/SFC	1.83	1.88	1.91	1.93	1.95	1.96	1.97	1.98
ΔT	40	50	60	70	80	90	100	110
$1/\Delta T$	0.025	0.02	0.0167	1.0143	0.0125	0.0111	0.01	0.0091

shown in Fig. 7-10. This quite clearly rectifies the data, allowing for reading uncertainties.

The equation of this line is

$$\frac{1}{SFC} = \frac{k}{\Delta T} + b$$

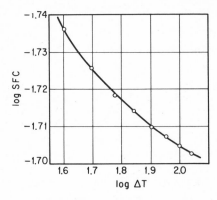

Figure 7-9 A log–log plot of the data for Example 7-3 showing that such a transformation fails to rectify the data. For such a small range of variables, log–log plots are best made as shown here rather than using commercial paper.

where b is the intercept at $1/\Delta T$ of zero axis and equal by inspection to 2.07; k, the slope of the line, is found graphically from Fig. 7-10 and is equal to -9.5. Then, from Eq. 7-7c,

$$\frac{1}{\text{SFC}} = 2.07 - \frac{9.5}{\Delta T}$$

or

$$\frac{\Delta T}{2.07\,\Delta T - 9.5} = \text{SFC}$$

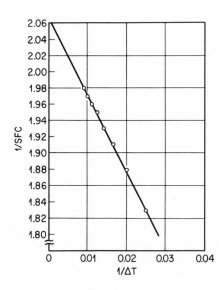

Figure 7-10 Data for Example 7-3 plotted as reciprocals, a transformation that yields a straight line.

the familiar hyperbolic form of the equation. The data in the form of Fig. 7-10 and the equation derived from it are quite revealing. For example, we see that even if the engine were run at a very high temperature, the SFC would drop to only 0.485, just slightly below that already achieved in the test. Extrapolating to the other end, it appears that an SFC of infinity will result when ΔT drops to 4.7°F. This probably means nothing more than that such an extrapolation is much too big to give any valid information.

The seasoned experimenter might, at this point, wonder if we did not cheat a bit by assuming that $T - 80$ rather than T is the significant variable. Should we not be able to find this fact in the data rather than from theoretical considerations? Let us see how this might be done.

From Fig. 7-10 we see that the equation involving the variables in their "raw" form is

$$\frac{1}{\text{SFC}} = b - \frac{k}{T - a}$$

This transforms to

$$T = \frac{k\,\text{SFC}}{b\,\text{SFC} - 1} + a$$

which is the form of Eq. (f) in Table 7-1 with T as Y and SFC as X. Let us tabulate the necessary quantities to perform the test for this function as follows, with X_1 equal to an SFC of 0.505 and Y_1 equal to the

SFC − 0.505		0.040	0.027	0.018	0.013	0.008	0.005	0.002
$T - 190$		−70	−60	−50	−40	−30	−20	−10
$\dfrac{\text{SFC} - 0.505}{T - 190} \times 10^4$		−5.72	−4.5	−3.6	−3.25	−2.68	−2.50	−2.0

corresponding T of 190°F. The plot of the third row versus SFC is shown in Fig. 7-11. As might be expected, the line is straight, indicating that this formula will correlate the basic data. From the usual slope–intercept method, we get from Fig. 7-11 the equation

$$\frac{\text{SFC} - 0.505}{T - 190} = 0.0046 - 0.0096\,\text{SFC}$$

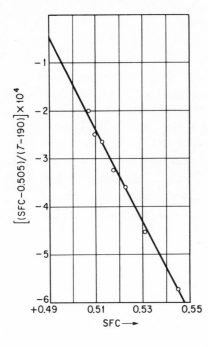

Figure 7-11 In this curve, an alternative form from Table 7-1 is tested with the data of Example 7-3.

which can be algebraically changed to the form (by solving for SFC)

$$\text{SFC} = \frac{T - 80}{2.09T - 157}$$

which is identical with Eq. (*f*) in Table 7-1 when this equation is put in the form

$$Y = \frac{T + a}{bT + ab + k}$$

Comparing these last two equations, we see that *a* is −80, *b* is 2.09, and *k* must equal 157 − *ab*, or −10. These constants in Eq. (*f*) in another form give finally

$$\frac{1}{\text{SFC}} = 2.09 - \frac{10}{T - 80}$$

What exactly have we proved in this analysis? Simply this: The figure for surrounding temperature did not have to be assumed or known, and it was not necessary even to recognize that $T - 80$ was significant. *This information exists and existed in the raw data themselves.*

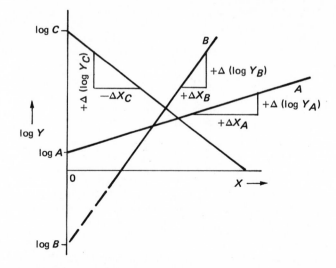

Figure 7-12 Graphical presentation of Eq. 7-8, the slope–intercept equation, applied to axes having only one zero, showing three variations.

If one axis of the rectified graph has no zero, such as the logarithmic scale or the probability scale shown in Chap. 2, it is usually possible to use the slope–intercept method if we extend one axis or the other. Figure 7-12 shows the three results of a rectification in which the Y axis is logarithmic. The equations are almost the same,

$$\text{Log } y_a = K_a x_a + \log a \tag{7-8a}$$

$$\text{Log } y_b = K_b x_b + \log b \tag{7-8b}$$

$$\text{Log } y_c = \log c - K_c x_c \tag{7-8c}$$

but we must extend the B curve below the x axis.

Example 7-4 A student, wishing to formulate rules of biological growth, programs a computer model in which a field of "food," 14 units on a side, has one or more "organisms" introduced into its center. The organisms have the value 10 and move randomly about in the food field assimilating (adding) the food of value 5. When an organism reaches a value 20, it splits and becomes two organisms of value 10 each and so on. After 33 moves (time intervals), the organism field is shown in Fig. 7-13 at which time 3 units of food remain to be eaten (not shown because the food field is a

0	10	10	0	0	10	15	0	0	0	15	0	0	15
0	0	10	15	0	0	15	15	0	0	0	0	10	10
0	15	0	15	0	0	10	0	10	0	0	10	0	15
0	15	0	0	0	0	0	10	15	0	15	10	0	0
10	0	15	15	0	0	15	15	0	0	15	0	0	15
10	0	0	10	0	0	0	0	0	10	0	10	0	0
0	10	15	10	0	15	0	0	10	15	0	0	10	10
0	0	15	0	0	10	0	0	0	10	0	15	0	0
15	0	15	0	10	0	0	10	15	0	0	0	0	15
0	0	10	15	0	0	0	0	0	10	0	10	0	10
10	0	0	0	0	15	10	0	0	0	10	15	0	0
15	0	10	0	0	10	0	10	0	0	0	10	0	10
15	10	0	10	0	15	15	0	0	0	10	0	10	0
15	0	0	10	0	0	0	10	0	10	0	15	0	0

Figure 7-13 After 33 "time intervals" the simulated "organisms" occupy the food field of the Example 7-4 computer experiment. Note that some organisms are "half-grown"; that is, they have the value 15, showing that they have assimilated 1 food unit.

separate matrix). This is what is known as a *computer simulation*, an experiment run on a computer.* Figure 7-14 shows the growth of the number of individuals, starting with a single individual, until all the food is assimilated (35 time units). The student wishes to develop equations that will predict the behavior of this, admittedly, simplistic model.

Solution In Sec. 4-6 we saw how mathematical analysis could assist in dimensional analysis. In this example, graphical and mathematical analysis

*Schenck, 1970, pp. 99–116.

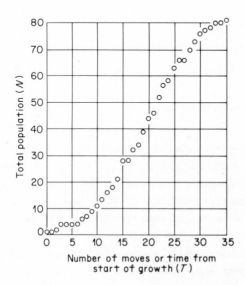

Figure 7-14 The growth of population in the Example 7-4 computer growth simulation as a function of "time."

share the spotlight equally. We begin by attempting a mathematical analysis, first defining our variables:

A = number of food particles to make an organism

F = number of food particles at start of growth ($14 \times 14 = 196$)

K = growth-rate proportionality factor

N = number of organisms at any time T

S = number of organisms at start when time is zero

$A(N-S)$ = food used to create $N-S$ organisms

$C = F + AS$ = the total "organic" material or "biomass" in the "experiment"

A very reasonable growth law for this sort of system can be stated as follows: The rate of growth dN/dT is directly proportional to the number of organisms N and to the amount of food remaining, $F - A(N-S)$. Thus, introducing C from the table,

$$\frac{dN}{dT} = KN(C - AN) \qquad (7\text{-}9)$$

To learn how this can help us plot the data, we must first integrate over the limits 1.0 to N for N, and zero to T for T. This gives

$$\ln \frac{CN - AN}{C - AN} = KCT \qquad (7\text{-}10)$$

Evidently, if we plot T on a linear axis and the "dimensionless growth parameter" $(CN - AN)/(C - AN)$ on a log axis, we shall obtain a straight line. Referring to the table, we saw that C was the total biomass of the experiment and equal to 196 food units, plus the equivalent, 2 food units in the first organism, or 198. A, at first glance, should be 2.0, because it takes two 5.0-unit "feedings" to construct a new individual. This is only true, however, in the infinite food field. In Fig. 7-13, we note that there are 80 organisms that have consumed 195 food units (3 remained when Fig. 7-13 was obtained). Thus A, in this restricted space, is $195/80 = 2.44$, or approximately 2.5. This makes an important point that will be stressed again in the next section, namely, that *constants in fitting equations should be obtained from, or at least be consistent with, the data themselves.*

Using these values, we can now plot Fig. 7-15, and we see that within the natural variation of the data, a reasonably straight line results. We can find the slope KC by noting that $(\ln 100 - \ln 10)/(20-9) = 0.21$, and the intercept at $T = $ zero is 1.5. Then Eq. 7-8*b* gives

$$\ln \frac{CN - AN}{C - AN} = 0.21T + \ln 1.5$$

Now the intercept of this line is of no special importance, because it is strongly influenced by how rapidly the organism moves out into the food

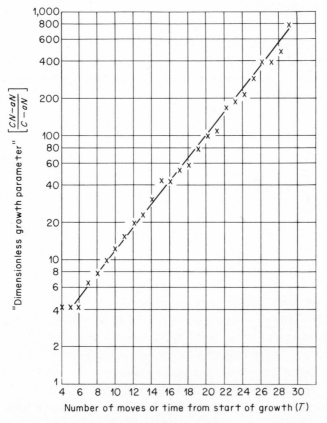

Figure 7-15 A dimensionless growth parameter found by theoretical reasoning plotted logarithmically against "time," for the data of Fig. 7-14, showing that rectification has been achieved.

field, which is a random matter; but the slope is of considerable interest. Suppose, for example, that we now made the organisms find several kinds of food before they could mature or gave them more food during growth or made some other change. Then the comparison of the slope value for the new experiment with this one would suggest how large an effect the change produced.

Another parameter of interest is the value of N at which the growing population is affected by the finite extent of the food grid, that is, the point at which "overpopulation" sets in. The shape of Fig. 7-14 suggests that a test of the second derivative will locate the *inflection point,* that point at which the change in growth rate goes from a positive to a negative value. Equation 7-9 shows the first derivative of the function. If we form the second derivative dT^2/dN^2 and set it equal to zero, we find a root,

$$N = \frac{C}{2A} \tag{7-11}$$

so that for this experiment the point at which the food field extent influences the growth rate is an N of $198/(2 \times 2.5) = 39.6$, which, Fig. 7-14 shows, occurred about 19 time intervals from the start.

Does this mean that bacteria in a culture or fingerling trout in a tank will act the same way? Not a bit. But the simulation does give us some ideas about how such systems *might* behave.

Notice the methodology of this example. First we noted that the data was reasonably precise (Fig. 7-14). Then we attempted to find a theoretical model of the system. Using this, we discovered a dimensionless growth parameter that allowed us to rectify the data and thereby verify the analysis (Fig. 7-15). From the plot of the rectified data, we obtained constants of interest, such as the slope. Finally, we used the mathematical model to obtain another important characteristic of the system, how its "extent" affected growth. Although this book is organized with particular techniques appearing at particular places in the experimental sequence, real-world experiments are much more akin to a musical fugue. That is, they progress, now depending on one strong theme, now another, now back to the first, until the entire body of data and analysis forms a harmonious whole.

When data are rectified on coordinates that do not have zero on either axis, the slope–intercept equation becomes a *slope-point equation.* When, for example, data are rectified on log–log coordinates, we know that Eq. 7-1 fits the data, and Eq. 7-2 gives the equation of the rectified line. We find the exponent a from measurements on the slope of the line:

$$a = \frac{\log Y_1 - \log Y_2}{\log X_1 - \log X_2} \qquad (7\text{-}12)$$

To find $\log k$, we must select a point *on the line and not necessarily a data point* and solve Eq. 7-1:

$$k = \frac{Y}{X^a} \qquad (7\text{-}13)$$

Example 7-5 What is the function relating the three lowest L/D points in Fig. 7-5?

Solution The slope of the line is found from Eq. 7-12, reading from the graph:

$$a = \frac{\log_{10} 0.043 - \log_{10} 0.009}{\log_{10} 34 - \log_{10} 8} = 1.071$$

As we noted before, we expect a fluid in steady flow to have a function, $\Delta P = K(L/D)^1$ (see Eq. 4-1). We now choose a point on the line, $\Delta h = 0.02$ in H_2O or 0.104 lb/ft^2 and $L/D = 17.0$. Then Eq. 7-13 shows that

$$k = \frac{0.104}{17^{1.07}} = 0.00501$$

Then Eq. 7-1 gives the equation

$$\Delta P = 0.00501 \left(\frac{L}{D}\right)^{1.071}$$

Referring again to Eq. 4-1, we note that if this L/D region is following the Fanning law, k should be $(f\rho V^2)/2$. The maximum exhale velocity in the experiment of Fig. 7-3 was estimated as 19 ft/s. Air density is 0.08 lb$_m$/ft^3, and at the estimated Reynolds number of 7150, the large friction factor f is 0.019. Then our expected k can be found as

$$k = \frac{0.019 \times 0.08 \times 19^2}{2 \times 32.2} = 0.0085$$

This is somewhat more than the experimental value but perhaps not too bad for three points. In fact, the experimental value is probably better. We used a maximum estimated velocity of 19 ft/s to compute a theoretical k value of 0.0085. If we had used the mean value of tube air velocity, 12.1 ft/s, then k would be 0.0035, much lower than the experimental value. Thus we expect that our problem is more in the selection of the correct Fanning velocity than in deficiencies in the data.

We thus conclude that the three lower points on Fig. 7-4 do indeed form a separate region within which the exhaled air behaves as a steadily flowing, frictional gas. Two pieces of evidence confirm this conclusion: the value of 1.07 for the exponent of L/D, compared to a theoretical expectation of 1.0, and the good comparison between the value of the experimental slope k and the expectations from theory and previous experimental work.

7-3 EXTRAPOLATION

One of the important reasons for rectifying test data is to enable us to *extrapolate* it to both ends of its extent. There are two important reasons for extrapolating test data: One is that the extrapolation may expose bad data by the failure of the data set to extrapolate to the expected value as one function reaches zero or infinity. The other is that the extrapolation may obtain for us a constant of the experiment or experimental function. We have already done this in obtaining the intercept point for the slope–intercept method of obtaining the equation of rectified data. This, however, was really a convenience rather than a necessity, since we could as well have taken any other point on the line and obtained the equation.

Rectification essentially involves separating one variable into Y and the other into X such that

$$Y = AX + B$$

Thus in *every* rectified set of data, we can plot Y versus X and find B when X is zero. If the rectified line is on log–log coordinates, we simply find k and a in Eq. 7-1 and plot Y versus X^a on regular graph paper. If Eq. 7-1 is the governing equation, such a line should pass through (0, 0).

Every function in X-Y space has two ends. Checking the zero end is readily done, as we have already shown and discussed. To check the "infinity" end, we simply plot $1/X$ versus $1/Y$ from our rectified function. If the data are continuously increasing in X and Y, we expect this plot to extrapolate to (0, 0). Sometimes plotting $1/X$ versus Y or X versus $1/Y$ will reveal useful information.

Although the shapes and magnitudes of experimental curves are often not known before testing or analysis are complete, it is often possible to estimate their terminal points through theoretical or common-sense reasoning. Thus the efficiency of any machine under zero load is *always zero*, no matter what value it may take in other situations. We know that when an alloy or mixture is varied as to components A and B, its properties will always extrapolate to the known properties of A as the percentage of B decreases and to the properties of pure B as the amount of A is reduced. In any type of flow-measuring device depending on head measurement, head versus flow must pass through the origin. In turbulent flow, the heat-transfer coefficient h is often a function of the mass flow per unit area G to the 0.8 power. If we plot $1/h$ versus $1/G^{0.8}$, we have a curve (if all is well) that should pass through the origin. When Ohm's law applies, the current is zero when the voltage is zero; whereas a plot of viscosity^{-1} versus temperature for a liquid should pass through zero at the liquid's freezing point. In spite of the uncertainties and potential errors involved in any extrapolation, this is often a good method for checking the overall consistency of numbers of data. Bad data may still extrapolate to the correct point, but failure to show a proper extrapolation will at least raise suspicions.

The main danger in extrapolation lies in the potential failure to recognize changing curvature beyond the range of the data. For example, Fig. 7-16 shows the results of tests on a solar cell bank in which progressively thicker Plexiglas is used to cover the cells and the resulting decrease in power noted. At first glance, we might expect the curve to extrapolate to zero power decrease at zero layers, when instead it extrapolates to about 3%. Actually, the moment a solar cell is covered by anything, even something with a perfect transmission characteristic, the cell bank gets warmer because of the loss of convection cooling and its

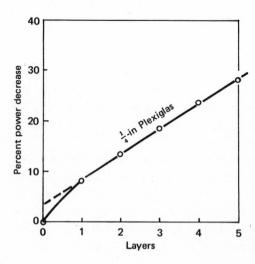

Figure 7-16 A linear-axis plot of the percent loss in power of a silicon solar cell bank as a function of the number of layers of $\frac{1}{4}$-in Plexiglas. The extrapolation to above zero does not, as explained in the text, indicate errors in the data, but such an extrapolation should always call for an explanation.

efficiency of converting solar rays to electric current decreases. The intercept of 3% at zero covering, far from indicating error problems, tells us that the bank under these conditions will show a 3% decrease when convection cooling is inhibited.

Example 7-6 A weir box having a triangular notch is calibrated by groups A and B. The following data result:

	Group A							
Flow (lb/s)	0.228	0.540	0.890	1.54	2.07	3.00		
Head (in over notch)	0.557	0.833	1.197	1.548	1.792	2.047		
	Group B							
Flow (lb/s)	0.054	0.822	1.40	1.82	1.95	2.78	3.95	6.25
Head (in over notch)	0.418	1.349	1.691	1.966	2.060	2.292	2.630	2.896

We suspect that at least one of these data sets is in error, because the group B data sheet indicates a zero reading for the hook gauge of 7.090 in (the scale reading as flow just starts) and the group A data sheet shows 7.293 in for the same quantity. Which (if either) of the two sets of readings is true, and how dependable is the better set?

Solution We know that head H and flow Q in a weir of any sort follow the general relation $Q = kh^a$ and that at zero head we must have zero flow. Thus an extrapolation to the origin $(0, 0)$ is the obvious way of checking these data. We first plot the two data sets on log–log paper to obtain the value of the exponent a and to enable us to transform the data into straight-line form. This is shown in Fig. 7-17. From these curves we see that the relationship for either set is $Q \approx h^{2.2}$, and we can now tabulate $h^{2.2}$, and plot it versus Q on linear coordinates. This is done in Fig. 7-18, and we see that group A's data fail to pass through the origin, whereas group B's extrapolation appears to come very close to $(0, 0)$. To bring the group A curve up so that its extrapolation is correct, we would have to add $(\Delta h)^{2.2} \approx 0.10$ or $\Delta h \approx 0.30$ in to all group A data. We noted that the group A reading is about 0.2 in higher than the group B reading; so if the group B data are true, the group A data are about 0.2 in too small in all head

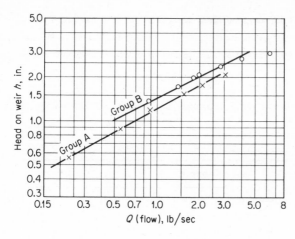

Figure 7-17 A log–log plot of the weir data of Example 7-6 showing that each data set has about the same exponent but that the two are displaced from each other.

readings, as predicted by our extrapolation. We can thus say with some assurance that the group *B* data are the preferred set. An investigator might be satisfied at this point to accept the group *B* data without further analysis. Interestingly enough, these actual data from two student laboratory reports reveal another and somewhat more subtle error of head measurement, as we shall now prove.

We would expect that if the head on a weir is increased to very large values, the flow will correspondingly increase and that infinite head and infinite flow will occur together. We can check this other end of the curve

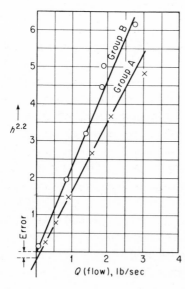

Figure 7-18 A plot of the rectified data of Example 7-6 using the exponent found in the plot of Fig. 7-17 to test the data for zero extrapolation. Clearly, the group *A* data do not pass this test.

by plotting reciprocals of Q and $h^{2.2}$. (If Q versus $h^{2.2}$ is a reasonably straight line on linear coordinates, then $1/Q$ versus $1/h^{2.2}$ will also be straight.) We expect, if all is well, that the curve will extrapolate through the origin. Let us first add $(7.293 - 7.090)$ or 0.203 in to each reading for group A, thereby correcting for the already established group A error in zero reading. Then we can tabulate:

Group A				
h (meas.) (in)	h (meas.) $+0.203$ in	$h^{2.2}_{correct}$	$(h^{2.2}_{correct})^{-1}$	Q^{-1} (s/lb)
0.557	0.760	0.5	2	4.4
0.883	1.086	1.2	0.83	1.85
1.197	1.400	2.1	0.476	1.12
1.548	1.751	3.4	0.295	0.65
1.792	1.995	4.57	0.22	0.485
2.047	2.25	5.9	0.17	0.333
Group B				
0.418		0.125	8.0	18.5
1.349		1.93	0.519	1.13
1.691		3.18	0.315	0.71
1.966		4.42	0.226	0.55
2.060		4.95	0.201	0.51
2.292		6.17	0.162	0.36
2.630		8.40	0.119	0.253
2.896		10.3	0.097	0.16

Figure 7-19 shows the plot of the last two columns in this tabulation. Although the resulting curve has some curvature at its lower end, the extrapolation is clearly not to the origin. Instead, this plot indicates that the flow will reach infinity at some finite head, a physical impossibility. Thus the data from both the group A and B determinations are defective, and we should logically suspect the head readings, which are usually more difficult to measure accurately than the flow rate. Investigation of the particular weir box in question revealed the difficulty and its cause. The hook gauge used to measure head was located so close to the outfall of the weir that at high rates of flow it read something less than true head, and this error grew progressively larger as flow increased, as shown in Fig. 7-20. The final result was, as shown in Fig. 7-19, a progressive error in h that had the cumulative effect of giving an improper zero intercept. A little thought should convince the reader that adding larger and larger amounts to the apparent h readings

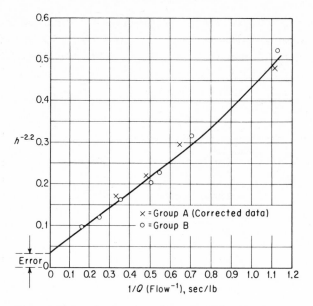

Figure 7-19 By plotting the reciprocals of the data shown in Fig. 7-18, we demonstrate that the data do not extrapolate to infinite flow at infinite head.

will indeed straighten out the $1/h^{2.2}$-versus-$1/Q$ curve and cause it to pass through, or at least closer to, the origin.

Example 7-7 Figure 3-4 shows a heat-exchanger test from which data and equations relating to errors in the test were discussed in Example 3-4. As the data table in Example 3-4 shows, the purpose of this test is to study the effect of increasing the hot-water rate in the inner tube on the heat-transfer

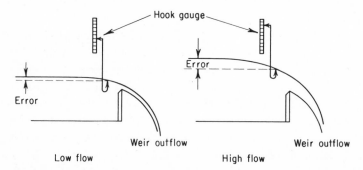

Figure 7-20 This diagram reveals the source of error in the weir test of Example 7-6. As the flow increases, the water surface curvature increases. This in turn causes the hook gauge, located too close to the outfall, to read depth in ever greater error.

rate between the hot and cold water. Continuing now with the actual test, we make a second run, points labeled B, and obtain the following "processed" A and B results:

Run number	Hot flow w_h (lb/min)	Reynolds no.	Heat-transfer coefficient U_i (Btu/h $\cdot$ °F $\cdot$ ft³)
1-A	0.7	900.	66.
1-B	0.85	1,080.	64.
2-A	1.0	1,280.	103.
2-B	1.19	1,500.	82.5
3-A	1.38	1.870.	108.
3-B	1.5	2,080.	105.
4-B	2.21	3,500.	128.
4-A	3.05	4,150.	160.
5-A	8.4	13,700.	243.
5-B	11.6	17,100.	264.
6-A	12.3	20,300.	283.

Can we obtain a relationship between the hot water flow rate w_h and the overall heat-transfer coefficient?

Solution We should really plot these data somehow to see if they are reasonably precise. Since we know that fluid and heat-transfer data is often rectified on log–log coordinates, we plot Fig. 7-21. It is tempting to interpret this graph as dividing into laminar and turbulent zones at N_{re} of 2000, and straight lines are drawn to suggest this. Also, we can obtain the slope of the line in the turbulent zone as 0.43. Such conclusions and analyses are premature. Figure 7-21 does *not* prove the function $U = kN_{re}^{0.43}$ but only suggests it. In fact, it is essential that we realize that the anticipated equations are

$$\frac{1}{U} = \frac{1}{H_i} + TR_0 \qquad (7\text{-}14)$$

$$H_i = kN_{re}^{0.8} \qquad (7\text{-}15)$$

where H_i is the turbulent hot water coefficient and TR_0 is the outside cold-water coefficient and is fixed because the cold water flow is fixed (see data sheet for Example 3-4). Thus before we can try the data in Eq. 7-15, we must *process* it using Eq. 7-14, and to do this we need the cold water thermal resistance TR_0. There are two ways to get TR_0. One is to estimate

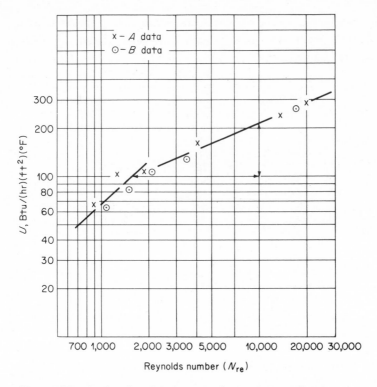

Figure 7-21 A log–log plot of the heat-transfer data in Example 7-7.

the conditions inside the outer annular tube and then apply a *correlation*, that is, an experimental equation, to *predict* the outside cold water coefficient. In this, and most similar cases, such an approach is incorrect. TR_0 *already exists within the data.* If we force a second value of TR_0 on the function, we both distort it and miss the chance of comparing our experimental TR_0 with that predicted from theory. Thus our second and correct approach is to obtain TR_0 by extrapolation.

Ideally, we would like simply to increase w_h without limit until $1/H_i$ is so small that TR_0 is the main quantity in Eq. 7-14. This suggests that we need an extrapolation to infinity, and because the high Reynolds number points emphasize the turbulent equation (7-15), we will plot $1/N_{re}^{0.8}$ versus $1/U$, as in Fig. 7-22. The plot shows that as $(1/N_{re}^{0.8})$ approaches zero, that is, as the hot velocity gets large without limit, the curve extrapolates to a U limit value of 400 Btu/h $\cdot$ °F $\cdot$ ft². Without going into details, the predicted annular resistance is found to be 385 Btu/h $\cdot$ °F $\cdot$ ft², an astonishingly good and heartening check, especially considering the scattered character of the data and the unimpressive energy balances observed in the data table of Example 3-4. However, it could have been much farther off without giving concern.

We now *process* the data again using Eq. 7-14 and a value for TR_0 of 1/400 and plot hot water H_i versus hot water N_{re} on Fig. 7-23, which summarizes the entire experiment and would be a most appropriate final graph in a report on this test. The line marked Eq. A is the prediction of the well-known correlation for turbulent heat transfer inside round tubes: The data are clearly beneath this, perhaps because of wall fouling (although the extrapolation should have included any fixed well resistance in TR_0), more likely because of corrosion blockages in the tube which would tend to give a product-type effect, that is, an effect that changes as the flow changes. This is partly confirmed by the extension of the data well below the normal transition zone, also suggesting turbulence promotion inside our supposedly smooth tube. Two possible terminations for our correlation line are suggested in Fig. 7-23, taken from heat-transfer texts. Notice especially that the rather vague suggestion that Fig. 7-21 gives two zones, changing at N_{re} of 2000, is not only spurious, but *the data show exactly the opposite*. Indeed, it is the continuation of turbulence below N_{re} of 2000 that gives the data a rather special aspect.

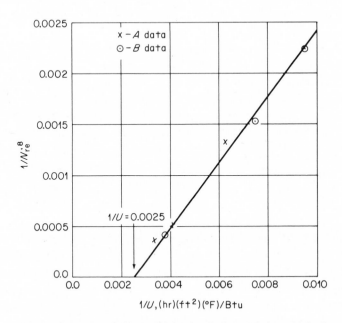

Figure 7-22 A plot of the data for Example 7-7 needed to obtain the cold water resistance by extrapolating to infinite hot water flow.

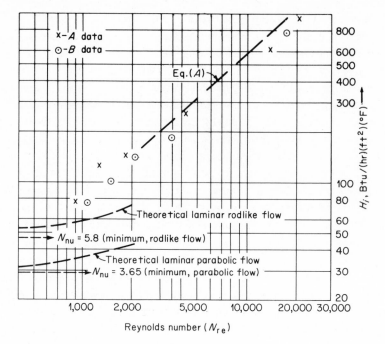

Figure 7-23 The computed heat-transfer coefficient for the hot water plotted against Reynolds number for the data of Example 7-7. The dashed line plots the theoretical equation:

$$H = \frac{0.023 C_p}{D N_{pr}^{2/3}} N_{re}^{0.8} \qquad (A)$$

Once again, we see a situation like that in Example 7-3 and Example 7-4, in which one or more constants of the experimental function could be found, and, in this example *must be found*, from the data itself. The ability to allow the data to "speak" fully, rather than forcing them to assimilate unpredictable, and often incorrect, theory, is what distinguishes the professional experimenter.

7-4 GRAPHICAL COMPARISON OF DATA

We have already seen one way in which two different sets of supposedly identical data can be compared in Examples 7-6 and 7-7, where we distinguished between two sets of data simply by using a separate symbol for each as we plotted the data to check and analyze it. In Fig. 7-17 it was obvious that the two data samples were drawn from different populations, and Fig. 7-18 showed the reason.

In Fig. 7-23 we see a suggestion that the A and B data may have some modest differences, suggesting that there was some difference in technique or in the test rig between the two tests.

These examples all involved simply plotting two experimental data sets Y_a and Y_b against a single X scale. There are several other ways to make data comparisons graphically when both data sets are precise enough to permit such interpretation.

Direct Plotting

The most obvious alternative plot to Fig. 7-17 or 7-18 might have been a direct plot of head on the weir for group A versus group B. Unfortunately, the two data sets were not taken at common Reynolds numbers. To get them on a common basis, we first need to rectify so as to enable us to interpolate between points, and by the time we have done this, we have available the very powerful extrapolation methods of Example 7-6, which solved the problems noted in the data. When data sets have been obtained with common X values, *cross-plotting* them, that is, plotting $Y_{a,x1}$ versus $Y_{b,x1}$, $Y_{a,x2}$ versus $Y_{b,x2}$, and so on, often reveals important facts.

Example 7-8 The student whose growth simulation model we considered in Example 7-4 decides to examine the effect of lowering the density of food, that is, the initial "biomass," in the food field by a factor of 2. He runs his second simulation with half as much food, 98 units, starts again with a single individual, and holds all else the same. The results show that the maximum population of 41 is about half that found in the previous simulation, 81, but

Population N	Time period T	Population N	Time period T
1	1	12	13
1	2	16	14
2	3	20	15
2	4	21	16
3	5	27	17
4	6	30	18
5	7	34	19
6	8	35	20
8	9	39	21
8	10	39	22
12	11	41	23
12	12	41	24

that the time to reach this number is not half that found before, but longer. What other comparisons can we make?

Solution We could, of course, treat this in the same way as the data in Example 7-4; that is, rectify, find the slope, and compare the data sets that way. A quick and in some ways more revealing plot occurs if we simply plot the N data from the full-food experiment against the same data for the half full-food experiment, as shown in Fig. 7-24. Notice that this is another version of the *parity graph* we discussed in Sec. 2-2. Now, without using anything more than a transparent straightedge, we can estimate the general trend of the line between an $N_{f/2}$ of about 8 up to $N_{f/2}$ of about 35, where the N_f data start to diverge from the completed growth pattern of $N_{f/2}$. What we see is that having the food field full gives the N_f growth a relatively immediate advantage of about five individuals, but that this advantage does not appear to increase until the $N_{f/2}$ growth is nearly ended. This is not an inherently obvious result. It might be more reasonable to expect that N_f would move continuously further from $N_{f/2}$ as T increased, and although Fig. 7-24 does suggest that this happens to some extent, it also suggests that the advantage is surprisingly small. Again, a crayfish farm might not show this behavior, but if it did, we would suddenly have a powerful computer tool by which to study it more deeply.

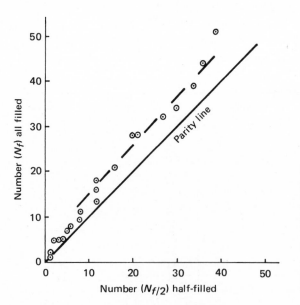

Figure 7-24 A plot of population for the full-food-field case against population for the half-filled field from the data of Examples 7-4 and 7-8. The parity line would hold if there were no difference in growth rate between the two growth simulations.

Nondimensional Plotting Using Averaging

Sometimes an investigator wishes to compare two sets of data that are completely unlike each other, except that they both are functions of the same X value. We might wish, for example, to compare dissolved oxygen in a stream at various randomly selected points with its fish counts taken at different points, or we might wish to compare frost depth at a location taken at various times over a year with deep groundwater temperature at the same location but not necessarily measured at the same times. Often the secret here is to average each data set on some consistent basis and then plot dimensionless ratios of the data $(Y_{a,1}/Y_{a,\text{avg}})$, $(Y_{a,2}/Y_{a,\text{avg}})$, . . . , and $(Y_{b,1}/Y_{b,\text{avg}})$, $(Y_{b,2}/Y_{b,\text{avg}})$, . . . , against X. This has the advantage of showing visually the relative location of the mean on the X axis where the line crosses the value of 1.0 and directly compares the two data sets on their functional behavior relative to their respective means.

Example 7-9 The sketch map of Narragansett Bay, Rhode Island (Fig. 7-25) shows the location of two sets of stations, those numbered 1 through 17 denoting locations used by the University of Rhode Island during the summer of 1971[*] and those with prefixes PR (Providence River), EP (East Passage), and WP (West Passage) referring to stations made by Morton[†] in the fall of 1965. The University of Rhode Island data show the mean value of K, the *attenuation function* for sunlight by the waters at each station during the summer. K is found by measuring the intensity of light I_0 and I_d at two points, D apart in depth, using the equation

$$\frac{I_d}{I_0} = e^{-KD}$$

Station no.	1	2	3	4	5	6	7	8	9
K_{avg} (ln/m)	0.626	0.524	0.271	0.423	0.546	0.607	0.725	0.817	0.679

Station no.	10	11	12	13	14	15	16	17	Mean value
K_{avg} (ln/m)	0.673	0.630	0.782	0.702	0.817	0.955	0.883	0.797	0.674

Morton's data show the mean concentration within the water column of the suspended material in the water in milligrams per liter. This is obtained by carefully filtering a large amount of water sample and weighing the residue:

[*]Schenck and Davis, 1973.
[†]Morton, 1967.

Station	PR-2	PR-3	EP-1	EP-2	EP-3	WP-1	WP-2	WP-3	WP-4	Mean value
Mean conc. (mg/liter)	4.71	3.59	2.155	1.825	0.86	2.745	3.005	2.685	1.47	2.561

We expect that there should be some correlation or relationship between these two parameters. How can this be shown, or not shown, with the given data?

Solution At first glance, we seem to have a real problem in that the stations are distributed in two-dimensional space. However, if you look at Fig. 7-25, you will note that there are really two lines of University of Rhode Island stations running north to south, those in the East Passage,

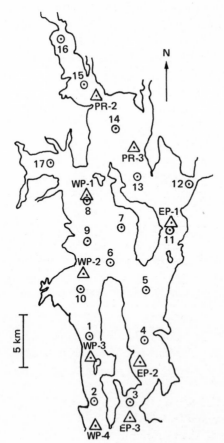

Figure 7-25 Sketch map of Narragansett Bay, Rhode Island, showing the location of stations noted in Example 7-9.

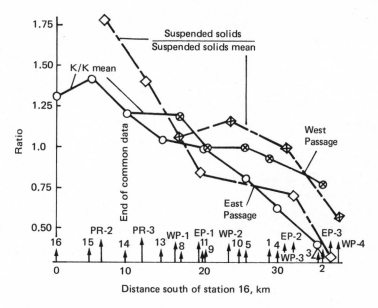

Figure 7-26 Nondimensional comparison of suspended-solid measurements and light-attenuation measurements along a north–south line on Fig. 7-25, but by two routes.

numbered 13, 11, 5, 4, and 3. Mortion's stations show a similar *bifurcation*, those with the WP prefix being in the West Passage and those labeled EP being in the East Passage. Stations 16, 15, and 14, and PR-2 and PR-3, are common to each side. If we form the ratio of each parameter value to the mean for the *entire estuary* and plot the East and West Passage data with separate symbols, but all against the number of kilometers south of station 16, we obtain Fig. 7-26.

This presentation shows two very important things: Both the suspended-solid parameter and K decrease over the entire north–south distance of the estuary, and both these parameters show a more rapid decrease in the East than in the West Passage. We have thus managed to compare "apples" to "oranges" reasonably convincingly. The next step might be to measure these two parameters at the same time and place and see if a stronger correlation could be achieved than appears in Fig. 7-26.

Comparing Two Sides of a Peaked Function

Occasionally an experimentalist wishes to compare the two sides of an experimental function having a single peak. Experiments involving the heating and then cooling of a system, the intake and outflow of the tidal prism in a bay, or the growth and decline of traffic around the rush hour yield data that is typically peaked with a growth portion followed by a decline. Often we wish to compare these two curves directly, although they may often follow quite different laws. If the central, or highest point, of the curve (Y_0, X_0), can be obtained by inspection, then one way to plot such data is to form the ratio Y_a/Y_0 with the right-side, or A data, and Y_b/Y_0 with the left-side, or B data. This gives a convenient 0-to-1.00 scale in the Y direction. In the X direction, form the absolute value $X_0 - X_a$ of the A data to the left of X_0 and $X_0 - X_b$ for the B data to the right of X_0. Either of these new X and Y variables may require rectification by using special coordinates or by mathematically altering the parameters as discussed in Sec. 7-1. With this simple transformation, it is often possible to rectify both curves and to obtain separate constants or equations characterizing the separate growth and decay of the Y parameter.

Example 7-10 Figure 7-27 shows the 4-year totals of sperm oil returned by U.S. whaling ships during the years 1806–1890. Points A and B refer to the War of 1812 period, during which relatively few ships were able to complete voyages. Point C refers to a similar situation during the U.S. Civil War when Confederate commerce raiding and other problems decreased the landings of oil. We wish to compare the growth and decline periods of this industry.

Solution Figures 7-28, 7-29, and 7-30 show three correlation attempts based on the suggestion of forming V/V_{max} versus some function of $T_0 - T$. Figure 7-28 fails to rectify either side. Figure 7-29 appears to rectify the growth side reasonably well. Notice how the growth line in this presentation defined by the points A and B returns rapidly after the wartime period to

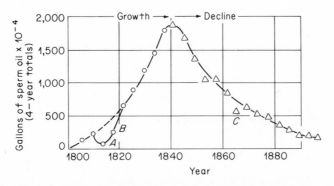

Figure 7-27 Four-year sperm oil totals plotted against the center data of the period as noted in Example 7-10. Points A and B refer to the effects of the War of 1812, point C to the effects of the Civil War.

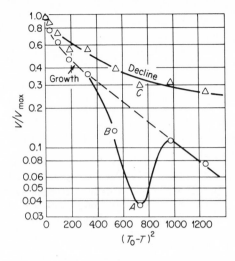

Figure 7-28 A correlation attempt based on taking the peak value at the year 1840.

the growth line established by the prewar points. Figure 7-30 appears to correlate most of the decline data and the five points in the last 20 years of the growth period as well. The advantage of restricting ourselves to Fig. 7-30 for both sets of data is that we can obtain a direct comparison of the two equations, although we can no longer include in the analysis the earlier growth data, some of which is widely deviating anyway. The general equation for both curves in Fig. 7-30 is Eq. 7-8c, but by forcing the functions to end at 1.0 when $T_0 - T$ is zero, we conveniently make log c equal zero. Thus, for either line,

$$\ln \frac{V}{V_{\max}} = -K(T_0 - T) \tag{7-16}$$

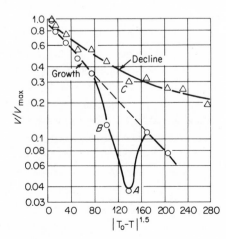

Figure 7-29 A second correlation attempt of the whale data of Fig. 7-27.

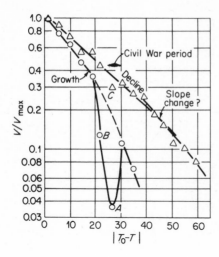

Figure 7-30 A third correlation attempt of the whale data of Example 7-10 showing reasonable rectification around the peak period.

where K represents the slope of the semilog plot. If we differentiate the above, we obtain

$$\frac{dV}{dT} = -KV \qquad (7\text{-}17)$$

where we have kept the negative sign on K through the differentiation. K for the growth period 1820–1840 is then 0.0552, and K for the decline period is 0.0374, or 69%, of the growth rate. The really remarkable thing about Fig. 7-30 is the sharpness of the peak between the growth and decline phases and the lack of any discernable inflection points. Let us see if we can use our discovered functions to propose a model of this industry. We note from Eq. 7-17 that the growth rate is proportional to V, the present "take" of the industry. This law is reasonable for an industry that has an essentially open demand and open resource base and is limited mainly by capital formation and possibly by new additions to the labor force. In a sense, Eq. 7-17 shows that some fixed fraction of the take went back in to expand the industry during the period 1820–1840. But now consider Fig. 7-31, which suggests what may have been happening around 1820–1840. The fishery was probably already depleting whales at some exponential rate corresponding to the growth exponential of the industry itself. This was in turn reflected by a proportional downturn in new whale production. At some critical point, probably during this time period, the yearly kills exceeded the new production. From this point on, as Fig. 7-31 suggests, each new increase in the industry had the double effect of reducing the whale stocks along the old curve, and then drastically reducing them further by the rapidly expanding difference between the declining production curve and the industry growth curve. And this, in turn, reduced the whale production curve well below its previous trend, and so on. In effect, sometime before 1840 a negative feedback loop was initiated that immediately and irrevocably altered the character of the industry. It was changed from a fishing enterprise that existed in some stable relationship with its stocks (although this must have ceased some time before 1840) to what became essentially a mining enterprise in which the existing whale stocks were hunted almost to extinction. If we now consider Eq. 7-17 in its relationship to the decline part of the data, we can readily apply the model suggested above. If we imagine that there is a given reserve of sperm whales V hunted worldwide such that in a given time period ΔT, $KV\,\Delta T$ are taken, then in the next ΔT time period, there are only $V-KV$ and only $K(V-KV)\,\Delta T$ taken, and so on. In the second period, $V_a - KV_a$ is the new V_a', so that we have a law

$$\frac{\Delta V}{\Delta T} = -KV$$

where the minus sign is necessary because ΔV is negative with ΔT positive. This is simply Eq. 7-17 in finite-difference form, showing that the decline of the fishery can be adequately explained by simple and continuous over-fishing with no significant replenishment, the same cause that probably produced the sharp peak between the two phases of the industry. Similar

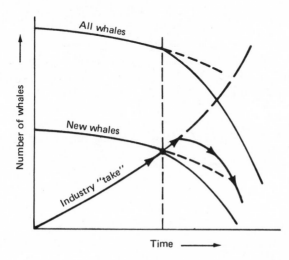

Figure 7-31 Sketch of the possible negative feedback system producing the sharp peak in sperm whale catches.

data sets are developing today in many aspects of world and national life. Their analysis and interpretation is no more trivial than the evident lessons of Fig. 7-27.

7-5 SUMMARY

Although this is a long chapter, the basic ideas are simple and readily summarized. When we have a set of *X-Y* data for study, we:

1. Plot *Y* versus *X* on some sensible coordinate system, often simply *linear* coordinates (Fig. 7-4). If the function shown therein is reasonably smooth, we tentatively assume that we can rectify it and find the mathematical function that describes its behavior. (Note that the data do *not* have to be "perfectly" smooth or without irregularities. Example 7-4 showed considerable experimental variability, and Example 7-7 considerable precision error. In each case we were still able to rectify and study the functions.)
2. Rectify the function as described in Sec. 7-1 and thereafter.
3. From the rectified line, obtain the mathematical function expressing it. If the experiment involves finding a property of a system, such as a friction factor or drag coefficient, it may only be necessary to obtain a slope or intercept from the curve, the governing function being already known.
4. Study the discovered function. This may give (a) other ways to plot the data to check the end points, (b) constants of the experiment or ways to obtain them, (c) insights into the meaning of the data through study of slopes, inflection points, or other aspects.
5. If you are comparing two functions having a common *X* parameter, the usual method is to plot them on the same rectified graph using different symbols. Other comparison methods involve direct plotting of the two sets, plotting the data in nondimensional form using an average or peak value in the denominator of *X* and *Y*, or dividing the function in half around its peak and starting the transformed *X* axis at this point.

Most of the chapter, of course, is not about these simple rules, but rather an attempt to show how these basic techniques interweave with the theory and mathematical character of the experiment to maximize our understanding of what the data are telling us.

PROBLEMS

7-1 How would you check the following experiments by extrapolation? Sketch the type of plot you would expect to get, and indicate how bad data might be revealed.

(a) A jackscrew gives a continuously increasing efficiency with load and follows the general formula $E = 100(1 - 1/L^n)$, where L is the load and n is some constant.

(b) A set of data on the viscosity of a fluid versus the temperature is given. The freezing and boiling points are known without error, and the viscosity of the vapor is one-thousandth that of the liquid at the pressure of the test. Show how both ends of the curve can be checked.

(c) Two different Orsat-type analyzers give percent CO_2 (unit A) and percent O_2 (unit B). The amount of CO is known to be zero. We check several samples of flue gas with each of the units in turn and obtain various percentages.

7-2 A test of a thermocouple gives the following data:

T (°C)	20	40	60	80	100
m (volts)	0.42	0.81	1.15	1.54	1.94

Are these data consistent?

7-3 A ventilating fan is operated at constant conditions, a series of probings is made at various distances from its outlet, and average velocity is obtained for each position.

Velocity (ft/min)	3000	1000	500	400	250	100
Distance of probe (in)	0	12	27	33	45	68

Are these data consistent?

7-4 A 116-V tungsten lamp is tested at reduced voltages. (It does not follow Ohm's law because of temperature effects.)

Volts	4	6	10	18	27	34
Amperes	0.0245	0.0370	0.057	0.0855	0.1125	0.1295

Are these data consistent?

7-5 Lead and zinc alloys are tested for melting point with various percentages of lead.

Percent of lead	50	60	70	80	90
Melting point (°C)	205	226	250	276	305

Are these data consistent?

7-6 Thorton's data show that the bacteria in a closed dish multiply according to the law

$$A = \frac{k_1}{10^{-k_2(T - k_3)} - 1.0}$$

where A is the plate area covered by bacteria at any time T after the start of the experiment and the ks are known constants except for k_1, which is unknown to the experimenter. How could data following this law be checked using straight-line extrapolation to known end points? If A_{total} is the area of the dish, what is its value in terms of the equation parameters?

7-7 Trevan's experiment with digitalis injected into frogs shows that the death rate D is related to the number of cubic centimeters injected x by

$$D = k_1 e^{-cx^2}$$

where k_1 and c are unknown constants. As in Prob. 7-6, show how data governed by this equation could be checked by straight-line extrapolations to known points or intercepts. In addition, give the expression using the test parameters noted for the number of frogs that will die during the test period of natural causes unrelated to digitalis injection and for the number that can withstand massive doses of the drug.

7-8 A pressure-measuring device is expected to follow the equation

$$\Delta P = K_1 H \frac{AB - K_2}{B}$$

where ΔP is the dependent variable or output; K_1 and K_2 are unknown constants; and A, B, and H are adjustable dimensions of the apparatus. What sort of straight-line extrapolation checks could be used for the following tests?

(a) P versus H with A and B held constant
(b) P versus A with H and B held constant
(c) P versus B with H and A held constant

7-9 In the test of a pressurized container, the law

$$P = \frac{Ast}{r + t}$$

is assumed to govern the specimens, where P is the failure pressure, r is the inner radius before the test and known exactly, t is the wall thickness and independent variable, and s and A are unknown constants. How could data from such a test be checked by straight-line extrapolation?

7-10 The normal stress P due to a shrink-fit of a collar of measured inner diameter b, outer diameter c, and interference when cold i is

$$P = \frac{Ei}{kbc^2} (c^2 - b^2)$$

Explain how an experiment in which collars of several b values are tested and P is measured by strain gauges can be checked by extrapolation methods using straight lines if E is an average (but unknown) modulus of elasticity and k is an unknown fitting constant.

7-11 In Probs. 7-11 through 7-15, find the rectified line and equation. A cooling experiment gives:

Temperature drop, ΔT (°C)	19.9	18.9	16.9	14.9	12.9	10.9	8.9
Time from start, θ (min)	0	3.45	10.87	19.30	22.8	40.1	53.75

7-12 A nonlinear circuit device gives:

Volts	67.7	65.0	63.0	61.0	58.25	56.25
Amperes	2.46	2.97	3.45	3.96	4.97	5.97

7-13 A pipe fitting gives the following pressure-drop data:

Velocity (ft/s)	1.3	1.45	1.6	1.7	1.95	2.5
Pressure drop (mm Hg)	0.200	0.25	0.290	0.32	0.43	0.70

7-14 A water pump gives the following data (rpm constant):

Flow (gal/min)	0	2	4	6	8	10	12
Head (ft)	98	106	112	114	117	112	105

7-15 An alloy gives the following property data:

Percent of A in mix	0	0.2	0.4	0.6	0.8	1.0
Specific heat	0.2	0.27	0.35	0.47	0.69	1.0

[*Hint*: Try test for (g) in Table 7-1.]

7-16 In any or all of the above five problems, study the function you have found for end points, and note any other items or values of theoretical interest.

7-17 Prove that any or all of the tests (*a* through *g*) in Table 7-1 will rectify the data equation noted.

7-18 Find the best equation for the following straight-line data relating the fraction of velocity head remaining downstream to the percent flow area blocked off in a hydraulic fitting having adjustable blockage. In addition, decide whether the extrapolation(s) of these data is (are) reasonable.

Fraction remaining	0.86	0.85	0.60	0.55	0.52	0.45
Percent blockage	10	20	30	40	50	60

7-19 In the following experimental functions, decide whether inflection points and/or maxima or minima occur and, if so, how knowledge of such points could assist in the experimental operation or in the data analysis.

(*a*) The Francis formula for a contracted weir is $Q = 3.33(B - 0.2H)H^{3/2}$, where Q is the flow in cubic feet per second, H is the head on the weir in feet, and B is the width of the spillway. We wish to run a test on a specific weir to verify (or modify) the constant 3.33 in the equation.

(*b*) A correlation relating the heat-transfer factor J to Reynolds number and the ratio of tube length L to tube diameter D is

$$J = 0.023 \left(\frac{VDP}{\mu}\right)^{-0.2} \left[1 + \left(\frac{D}{L}\right)^{0.7}\right]$$

We wish to verify this equation for a special kind of entrance condition by holding everything constant but the tube diameter and finding the effect of D on J. Around what values of L/D should we operate our test?

(c) Richardson's equation for the electron emission current i from a surface heated to an absolute temperature T is $i = AT^2 \exp(-b_0/T)$, where A and b_0 are constants. Given a plot of i versus T, how could we estimate a value of b_0? (*Hint*: Investigate the slope changes using the second derivative.)

7-20 Referring to the simulation described in Examples 7-4 and 7-8, the data shown resulted from a simulation in which the food grid was entirely filled with number 5s, but the organisms had the ability to hunt for food in any of the four surrounding spaces. Analyze as in the examples.

Population N	1	2	3	4	5	6	7	8	9	10	11	12	13	14	15	16	17	18	19
Time period T	1	2	2	3	4	5	7	9	12	16	21	28	35	44	54	62	67	79	82

7-21 In another of the Example 7-4 simulations, the organisms are allowed to hunt for food as in the previous problem but must assemble five value 2 parts to produce a new organism. Analyze as in the examples.

Population N	1	1	1	1	2	2	2	2	2	3	4	4	4	4	5	7
Time period T	1	2	3	4	5	6	7	8	9	10	11	12	13	14	15	16

Population N	8	8	8	8	9	11	13	13	13	15	17	19	21	21	23
Time period T	17	18	19	20	21	22	23	24	25	26	27	28	29	30	31

7-22 The shape factor per unit drainage length S/L between a horizontal ground (source) plane of infinite extent and a vertical (sink) plane of height Y located $W - Y$ distance below the ground plane, where W is the total vertical extent of the system, is given by

Y/W	0.5	2.5	4.5	6.5	8.5
S/L	0.486	0.698	0.946	1.249	1.956

Rectify the data and obtain the value of S/L when Y/W is zero, which is the residual shape factor due to adjacent seepage fields.

7-23 A computer simulation of pollutant dispersion following a so-called random walk model gives the following mean radius of the dispersing cloud as a function of time. What law is suggested in the cloud growth?

Time	1	2	3	4	5	6	7	8	9	10
Cloud radius	1.57	1.85	2.19	2.26	2.48	2.54	2.70	3.18	3.28	3.46

Time	11	12	13	14	15	16	17	18	19	20
Cloud radius	3.70	3.94	4.18	4.41	4.41	4.67	4.68	4.54	4.82	4.62

The cloud should start with a radius of 1.0 and expand continuously. Check the end points.

7-24 A spherical buoy can be "tuned" by moving a weight up or down a vertical rod sticking out of the top of the buoy. The roll resonant frequency as a function of d, the distance of the weight from the top of the rod, is given by

Period (s)	3.27	3.48	3.68	3.86	4.23	4.47	4.97	5.53	6.16	7.05	8.90
d (cm)	25	23	21	19	17	15	13	11	9	7	5

Rectify and obtain the governing equation. What is the maximum wave period in seconds to which the buoy can be tuned to roll? At what d value will the buoy period be zero seconds?

7-25 The volume rate of flow through a small orifice is measured at various heads. The results are

Q (in^3/s)	1.11	1.10	1.09	1.07	1.01	1.00	0.978	0.94	0.90	0.85	0.774
Head (in)	5.69	5.44	5.19	4.94	4.69	4.44	4.19	3.94	3.69	3.44	3.19

Find the experimental function relating these two quantities. We expect orifices to behave the same way as the weir in Example 7-6. Make the same graphical checks.

7-26 The stress–strain behavior of an elastic shock cord is measured by loading the cord with weights and measuring the length:

Weight (kg)	0.143	1.83	0.223	0.263	0.303	0.343	0.383	0.423	0.463
Length (cm)	14.3	16.2	18.6	21.1	24.2	26.8	29.8	32.4	34.9

Metallic bodies follow Hooke's law (strain is proportional to stress), but rubber strands do not. What is the law relating these variables?

7-27 A silicon solar cell is tested under incandescent and then fluorescent light. In each test, the load resistor is varied to obtain an E-versus-I curve for the cell.

	Incandescent							
I (mA)	1.5	0.129	1.12	1.0	0.63	0.36	0.13	0
E (V)	0.15	0.60	1.11	1.48	1.75	1.85	1.96	2.04

	Fluorescent								
I (mA)	0.16	0.150	0.142	0.128	0.119	0.092	0.08	0.062	0.036	0.0
E (V)	0.02	0.25	0.38	0.50	0.566	0.75	0.87	1.0	1.12	1.25

The outputs are different because no effort was made to keep the lamp–cell distance the same. A good reference point when testing cells is the *maximum power* point, which can be found by plotting EI against E. If this is done, we find that the incandescent peak power occurs at $I = 0.99$ mA and 1.52 V and the fluorescent peak power point is at 0.11 mA and 0.65 V. Compare these two response curves with regard to their shapes and intercepts.

7-28 We wish to compare the power rise part with the power decline part of the EI curve for the incandescent and fluorescent lights in Prob. 7-27. Plot these EI-versus-E curves and decide how to compare the two sides of the functions for the two sides of the functions for the two kinds of light. Based on this study, do you feel that the two lights differ significantly with regard to solar cell response characteristics?

7-29 In a lobster-tagging experiment, offshore lobsters are placed close to shore and then their progress seaward checked with trap returns. The following data summary gives the rounded distance in miles at which tagged specimens were found at various times after the start of the experiment:

Days from start	16.	70	100	150	250
Maximum miles from start	5.5	11.0	14	17.2	23

Discover the law that governs this movement. These lobsters were taken from an offshore group 60 miles at sea. Predict the minimum time of their return to their starting point.

7-30 A model ground-effect machine with a 6-in skirt is loaded in 25-lb increments and its hover height measured.

Payload (lb)	0	25	50	75	100	125
Hover height (in)	3.2	2.5	2.0	1.6	1.2	0.95

Obtain the equation of this law. The unloaded weight of the machine is 105 lb. If we could take off 25 lb, what would be its new unloaded hover height?

7-31 In another experiment involving the ground-effect machine, the unloaded machine was tested with various skirt heights and its static hover height measured.

Skirt length (in)	0	1.5	3.0	4.0	5.0	6.0
Hover height (in)	0.4	0.5	0.75	1.25	2.2	3.25

When a ground-effect machine is operating far from the ground, its nozzles, which assist in maintaining an increased pressure underneath the machine, are ineffective, but as it gets closer to the ground, the nozzles on the periphery assist in holding pressure under the machine. See if you can find the height at which this mode of operation changed during this test.

7-32 A series of hydrographic maps gives the following relationships between the length L of a stream and the area A it drains.

L (km)	6.0	11.0	18.0	28	51	72
A (km^2)	7.1	21.1	48.0	115	305	480

Find the law relating to these two variables. At what length of stream are L and A numerically equal?

7-33 The dilution of secondary sewage is related to the discharge flow rate from the outfall as follows:

Dilution	40	35	30	25	20
Discharge flow $(gal/day \times 10^{-6})$	2.1	3.2	4.4	6.7	10.05

How are these quantities related? At what flow would there be zero dilution?

BIBLIOGRAPHY

Doeblin, E. O.: *Measurement Systems Application and Design,* McGraw-Hill, New York, 1975.

Doolittle, J. S.: *Mechanical Engineering Laboratory,* chap. 15, McGraw-Hill, New York, 1957.

Graham, A. R.: *An Introduction to Engineering Measurements,* Prentice-Hall, Englewood Cliffs, N.J., 1975.

Hoelscher, R., J. N. Arnold, and S. H. Pierce: *Graphic Aids in Engineering Computation,* chaps. 2 and 4, McGraw-Hill, New York, 1952.

Holman, J. P.: *Experimental Methods for Engineers,* McGraw-Hill, New York, 1971.

Mackey, C. O.: *Graphical Solutions,* chaps. 5 and 6, Wiley, New York, 1943.

Morton, R.: *Tech. Mem. 396,* Naval Underwater Weapons Station, Newport, R.I., 1967.

Oldenburger, R.: *Mathematical Engineering Analysis,* Macmillan, New York, 1950.

Schenck, H.: *Case Studies in Experimental Engineering,* McGraw-Hill, New York, 1970.

Schenck, H., and A. Davis: A Turbidity Survey of Narragansett Bay, *Ocean Engineering,* vol. 2, pp. 169–178, 1973.

Smith, E., M. Salkover, and H. Justice: *United Calculus,* Wiley, New York, 1947.

Wilson, E. B.: *An Introduction to Scientific Research,* chaps. 11 and 12, McGraw-Hill, New York, 1952.

Worthing, A. G., and J. Geffner: *Treatment of Experimental Data,* chap. 6, Wiley, New York, 1943.

EIGHT

GRAPHICAL AND STATISTICAL ANALYSIS: DATA OF LOW PRECISION

Ideally, we would like to design all our experiments so that they meet the high-precision requirements of Chap. 7, but it should be evident that this simply is not always possible. There are great classes of inherently erratic data: weather and all other naturally occurring measurements, data on human and animal interactions, data so difficult to obtain that no sum of money will improve their precision.

An inexperienced industrial engineer might plot the average number of pieces rejected per day against the amount of time a worker has been in the company employ and hope to obtain a well-behaved function that perhaps discloses that the rejection rate drops at the 0.43 exponent of time at the job. In fact, it is most likely that a set of data will be obtained in which some statistical method must be used to prove that *any* positive relationship exists between the two variables. What may be needed is a *scatter diagram* and some sort of *statistical correlation* method, and the engineer is unlikely to do more than assume a straight-line function for any tenuous relationship discovered.

This chapter deals with data that is too scattered, either because of its inherent nature or of instrument and test control deficiencies, to do more than suggest an X-versus-Y relationship, but that we still wish to plot and perhaps use to get a simple equation. In addition, as is always the case with variables that have a high random variation, we must provide our graphs and data output with added information on *dispersion* of the data points and the chance of finding a given point near to or far from our "best" line.

8-1 REJECTION OF OUTLYING POINTS

Engineering tests seldom deal with more than a few determinations at a given apparatus configuration. The finding of such quantities as orifice coefficient, thermal conductivity, stack-gas composition, and so on, is usually made with tests in which 5 to 10 experimental determinations are considered adequate. If even one of these points is very different from the others, a so-called outlier, flyer, or wild point, it will exert a large and possibly detrimental effect on the numerical results and on the conclusions flowing from those results.

Opinions on the treatment of such faulty observations vary,[*] but it is often felt that they should be thrown out if some criterion for rejection besides a hunch or a desire for symmetry can be found. Once a criterion is chosen, it must be applied to every test point, even those that look all right. If there is any doubt, the point stays in the data set.

In this regard, it is well to distinguish between *end points* and *center points,* because these must be viewed differently. In Fig. 8-1, point *A* is quite obviously an outlier, and assuming that we have a rejection criterion that so orders, we should throw it out. On the other hand, point *B* may not be a wild point at all but merely the beginning of a new portion of the curve (transition to turbulence, passing the yield point, change of phase, etc.), and to omit it will deny us the most significant portion of the test data. Even point *C* may be both correct and important. We cannot tell, in any case, without taking more data at low *X* values. Thus it seems to the writer that a reasonable compromise between those who would throw out everything that looks slightly off and those who would never drop a datum point is given in the following simple rule: *Outliers should be rejected, following a statistical criterion, only if they are center points.*

[*]For a concise statement of two divergent opinions, see Dumond, 1953.

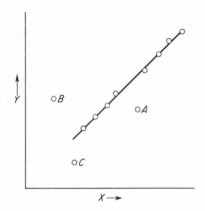

Figure 8-1 Illustrating some types of outliers. Point *A* is probably a bad point, whereas point *B* might be bad but could be the most important point of all. Since point *B* may represent a physical effect, point *C* should also be retained, at least until more data have been added to the plot.

Let us list some major criteria for data rejection of both a *physical* and a *statistical* nature:

1. *Obviously defective control.* In most tests we shall be taking a great deal of data that are not directly concerned with the machine or effect under investigation. Variables such as barometric pressure, line voltage, humidity, oil pressure, coolant flow, and so on, are not expected to exhibit any radical variation but are monitored just in case. Thus, if a wild point occurs at the same time that the data sheet shows a line-voltage surge, rejection of this point is entirely reasonable.

2. *Obvious instrument malfunction.* If the last several points taken on a given day or test series are all wild, and furthermore, they are distributed at random among the other points (as they will be if the test sequence is randomized), we expect that a measurement malfunction has occurred, and we should reject these points.

3. *Imbalance beyond a present criterion.* If a *balance equation* can be used to check each datum point, we can decide on some maximum amount of imbalance and reject all points that exceed it. Even this simple criterion must be applied with some care. In the data table of Example 3-4 we had heat-balance errors of almost 13%, but we showed by an error study that these were most likely the result of our inability to read the cold water temperature changes accurately enough. To reject these points would be unwise, because the cold water energy is used *only* to check the balance. If we do use a balance to reject, however, we must be consistent, regardless of whether the points look like outliers. All or none is the rule here.

4. *Exceeding a statistical criterion.* There are a number of ingenious statistical means of setting limits for point rejection.[*] We shall consider only one, *Chauvenet's criterion*, and we shall assume that our errors are normally distributed so that Table 2-2 can be used to find probability values. For other criteria and for nonnormal error distributions, the reader should consult the various sources listed in the Bibliography at the end of the chapter.

This rule states that any reading out of a series of n readings shall be rejected if the magnitude of its deviation from the true or mean value is such that the probability of occurrence of such deviation does not exceed $1/2n$.

The following table summarizes the requirements of Chauvenet's criterion for selected numbers of data (n):

No. of items (n)	4	5	6	10	15	25	50	100	300
Reject if T (Eq. 2-22) is more than	1.54	1.65	1.73	1.96	2.13	2.33	2.57	2.81	3.14

[*]See Proschan, 1953; Bennett and Franklin, 1954, pp. 665–668.

Rejection of points seems unwise when less than four data exist, particularly considering Fig. 8-1.

One disturbing thought usually occurs to engineers who find any idea of point rejection abhorrent. Suppose that one or two data are thrown out in following a statistical criterion, and a new standard deviation and probability limit are computed, which necessitates throwing out still more points. In theory, we might continue until almost all the data lie outside our acceptable limits. A rule of thumb to prevent such a disturbing (and almost unheard of) occurrence is to apply a statistical rejection criterion *only once*. Actually, if data are so strangely scattered as to produce this kind of mass rejection, statistical inferences should probably give way to a practical reevaluation of the test and instrumentation.

Example 8-1 An electrical flow of 70 mA alternating current passing through the heart may arrest it. To test the worst-case situation, 15 students were asked to clasp two metal pipes as tightly as possible and measure their hand-to-hand electrical resistance with a high-impedance ohmmeter. They then removed their right shoes and socks and found the resistance between an aluminum plate on which they put, with all their weight, their right foot and the pipe tightly grasped in the left hand. To motivate the problem, it was suggested that a toy train manufacturer has become concerned that his 16-V ac circuit might electrocute someone with a very low internal resistance and has queried a consulting engineer.

Subject	Hand-to-hand resistance (Ω)	Foot-to-hand resistance (Ω)
1	20,050	17,850
2	27,333	14,633
3	15,800	10,800
4	21,000	28,000
5	13,375	18,125
6	8,208	7,958
7	39,833	26,714
8	16,100	6,460
9	23,000	16,600
10	20,000	26,000
11	23,100	22,200
12	27,000	8,250
13	28,200	66,500
14	18,500	8,500
15	11,708	13,166

Our first thoughts should probably be of the usual statistical materials of Chap. 2. We certainly expect that both samples are drawn from a normally distributed population. What parameters might we look at?

Solution This actual data from a sophomore laboratory group at the University of Rhode Island was processed by the interactive computer program, STATPACK. We shall thus show and discuss a number of the items displayed by such programs and suggest how powerfully they allow us to manipulate and examine data. STATPACK does much more than we need here, and items of no interest will be omitted. We first obtain the so-called elementary statistics:

Variable	Total (Ω)	Mean (Ω^2)	Maximum (Ω)	Minimum (Ω)
Hand	313,207	20,880.465	39,833	8,208
Foot	291,756	19,450.398	66,500	6,460

Variable	Range (Ω)	Variance (Ω^2)	SD (Ω)	SD mean (Ω)
Hand	31,625	61,295,888	7,829.168	2,021.483
Foot	60,040	220,438,304	14,847.164	3,833.521

The total column is of use in many further statistical calculations. The maximum and minimum, and the range, suggest the dispersion of the sample, but the standard deviation and its square, the variance, does a better job. We could now proceed as in Sec. 2-8 and test the two samples for their normality, but with only 15 items, we can do almost as well by *inspection*. Viewing the hand data, we note that the mean is 20,880 Ω and that there are eight readings less and seven greater, a good check of the skewness. Now if we add and then subtract the SD for the foot data to and from the mean, we get a range of 13,051 to 28,709 Ω within which we expect a normal sample to have about two-thirds of its readings or 10 here. In fact, there are 12—not too good, but perhaps reasonable for only 15 data.

Applying the skewness test to the foot data, we find 10 data below the mean and only 5 above, a clear indication of problems or at least of a skewed distribution. But if we look at the foot column, we see at once an unusually high reading: Subject 13 showed 66,500 Ω. We might wonder if this very high reading did not bias the sample mean high, producing the skewness problem we noted.

To test subject 13's foot data, we can apply Chauvenet's criterion. We note that there are 15 data; so $1/2n$ is 0.0333. The chance of a point falling inside this is 0.9667 or, on one side, 0.4833. From Table 2-2, we see that T is 2.13 and we must reject a point if it is farther than 2.13 SDs from the mean. Thus,

$$19450 + 2.13 \times 14847 = 51070 \ \Omega$$

The foot reading of subject 13 is well beyond this limit; so we shall reject it. Since we wish later to *correlate* the two data sets, we shall also drop subject 13's hand-to-hand data. We can then *edit* our data to drop line 13 and now obtain for our new elementary statistics:

Variable	Total (Ω)	Mean (Ω)	Maximum (Ω)	Minimum (Ω)
Hand	285,007	20,357.641	39,833	8,208
Foot	225,256	16,089.711	38,000	6,460

Variable	Range (Ω)	Variance (Ω^2)	SD (Ω)	SD mean (Ω)
Hand	31,625	61,595,408	7,848.273	2,097.539
Foot	21,540	54,950,192	7,412.846	1,981.164

We removed only one set of the 15, but if you compare this new batch of statistics with the previous set, you find the following. The mean of the foot data, which was similar in size and slightly higher, now looks significantly less than the hand data. The standard deviation of the foot data, which was twice that of the hand data, now looks similar in size. The range of the foot is now two-thirds that of the hand rather than twice, as in the first data set.

But we might wonder about the high point of subject 7 in the hand column. Note that we must return to the *original data* to check this data point, because otherwise we defeat our rule about never throwing out data progressively. If we note as before that T is 2.13 for 15 data, then for the hand data we get an upper limit of

$$20880 + 2.13 \times 7829 = 37556 \ \Omega$$

by Chauvenet's criterion. Subject 7's value of 39,833 Ω is outside this range and should be dropped. Then our third set of elementary statistics, again with the entire row for subject 7 deleted to keep the two sides symmetrical for later *correlation*, becomes:

Variable	Total (Ω)	Mean (Ω)	Maximum (Ω)	Minimum (Ω)
Hand	245,174	18,859.535	27,333	8,208
Foot	198,542	15,272.461	28,000	6,460

Variable	Range (Ω)	Variance (Ω^2)	SD (Ω)	SD mean (Ω)
Hand	19,125	32,689,552	5,717.477	1,585.743
Foot	21,540	49,399,520	7,028.477	1,949.349

Comparing this result with the previous one, we see that the foot mean may well be significantly lower, a fact we shall consider in Chap. 9, but that the hand data may show less spread or variation, a fact not evident in the previous set of statistics. If we now check our foot data again, we note that there are six data above and seven below the mean, so the rejection of the two points has cured the skewness problem. If you now check both samples, you will find that no more points merit rejection. This essentially means that there is no other single datum point remaining in either sample that will drastically alter the statistical conclusions. What Chauvenet's criterion does is to single out exactly those data that are most seriously biasing the statistics obtained from the whole sample. This will become even more evident as we examine these data further.

We can now make an interim report to our toy train manufacturer: Noting first that the two points we rejected would lead to high results and thus to falsely safe conclusions, we see that from Ohm's law, a 16-V transformer will need an external resistance of $\frac{16}{0.07} = 228.6 \ \Omega$ or less to produce a lethal current. Looking at our third table we see that this is, for the hand case, a T of

$$\frac{18859 - 229}{5717.} = 3.26$$

Such a T value is well out in the tail beyond where 13 data points can tell us much. We know that we cannot have negative resistance anyway; so we might wonder about this distribution when it gets that close to zero. In any case, we can say that electrocution is unlikely, provided that the train set is not immersed in water.

The idea of assigning a limit to a deviation using a reasonable statistic has applications in experimentation beyond the simple rejection of wild points. In a variety of tests, such as stress–strain experiments using testing machines, a straight-line function is assumed to govern the test behavior up to a fixed but unknown limit. Detection of significant deviation from a straight-line law can be put on a firm basis using Chauvenet's criterion. A straight line through the data points (using least-squares methods as discussed next) will enable the experimenter to compute a standard deviation and, assuming that the deviations are normally distributed, assign a limit to the nth deviation. When the test indicates two points in a row that exceed the Chauvenet limit, the experimenter may logically assume that the limit of the straight-line law has been reached. Since the computation must be performed after each new datum point is added, it is best to exceed the straight-line region and then process the data, point by point, on a computer to find exactly where the deviation began.

8-2 THE LEAST–SQUARES METHOD OF STRAIGHT–LINE PLOTTING

The most accurate and most rigorous way of drawing a best line or *correlation line* among a set of points on an XY coordinate plane is the least-squares method. Suppose that we vary an independent and controlled variable X over its range or envelope and read from instruments a dependent variable Y. Suppose further that both Y and X have precision error, which is large at their low ranges and decreases as Y and X are increased. If we were to set X at all its possible values and repeatedly read Y until we had an infinite population of test points filling the (X, Y) envelope, we would have a two-dimensional normal distribution made up of an infinity of Y-reading populations. Plotting in the usual manner the number of readings at any interval versus the deviations, we might obtain the distribution sketched in Fig. 8-2.

In Sec. 2-6, we showed that when an instrument deviates in a normal manner, the sum of the squares of the deviations of the instrument from its best value must be a minimum. The same principles and proof will apply to the general case expressed in Fig. 8-2. Here, we have an infinite manifold of individual normal-type curves, to each of which the discussion of Sec. 2-6 will apply. Thus we can say the best line through a set of scattered points on an XY plane *must take the position that makes the sum of the squares of the point deviations from the line a minimum.* It is from this rule that the term "least squares" derives.

Let us consider the so-called classical least-squares problem, in which we know (1) that the infinite population of points in the XY plane delineates a straight line, (2) that all precision error is concentrated in the Y variable, and (3) that the precision error in the Y direction is the same for all Y values. With these

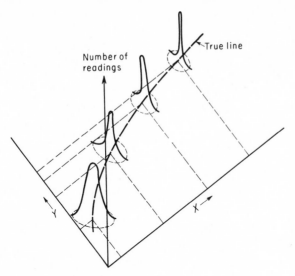

Figure 8-2 If we were to replicate many times four XY readings that were expected to fall on the true line as shown and if both the X and Y readings had uncertainty or precision error that decreased with increasing X and Y, we would build up "mounds" of readings as shown here. Such mounds are two-dimensional normal curves.

three substantial restrictions, the general manifold shown in Fig. 8-2 becomes the much less general manifold shown in Fig. 8-3. The reader might wonder if we have not so compromised our method as to make it almost useless. In Chaps. 2 and 3, we saw that it was quite typical for a dependent measured variable to have a changing precision error within different portions of its envelope. Often

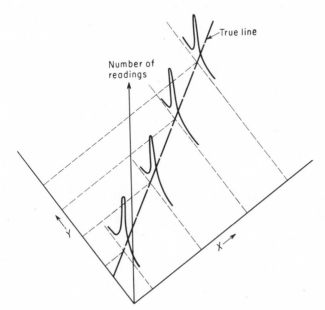

Figure 8-3 If we were to replicate many times four XY readings that were expected to fall on a straight line as shown and if only the Y readings had precision error that was constant for all Y values, we would build up the uniform manifold of points shown here.

we cannot say that we have chosen an X variable that is free of precision error, while the necessity of knowing that a straight-line function exists is hardly a commonplace of general experimental work. Yet to generalize this model of our XY function so adds to the computations and to the complexity of the interpretations that we are often tempted to apply our classical model to data that seriously deviate from the ideal. In many cases, however, it is possible through algebraic transformation to make our data tractable and reasonably approximate in form to this basic model.

Hoping that we can accept the model of Fig. 8-3, we now assume a general straight-line equation

$$Y_c = aX + b \qquad (8\text{-}1)$$

and we wish to obtain expressions for a and b such that the summed squares of the Y deviations from this line are minimized. Let Y_c be the true Y value at any X so that $Y - Y_c$ is the deviation at any X. We wish to minimize $\Sigma(Y - Y_c)^2$, which is the same as $\Sigma(Y - aX - b)^2$. Then

$$\frac{\partial \, \Sigma(Y - aX - b)^2}{\partial b} = 0 \qquad (8\text{-}2)$$

and

$$\frac{\partial \, \Sigma(Y - aX - b)^2}{\partial a} = 0 \qquad (8\text{-}3)$$

must hold true. If there are n readings, Eq. 8-2 becomes (because $\Sigma \, b = nb$)

$$nb + a \, \Sigma \, X = \Sigma \, Y \qquad (8\text{-}4)$$

and Eq. 8-3 becomes

$$b \, \Sigma \, X + a \, \Sigma \, X^2 = \Sigma \, XY \qquad (8\text{-}5)$$

We wish to find a and b so that a simultaneous solution must be made giving

$$b = \frac{\Sigma \, X^2 \, \Sigma \, Y - \Sigma \, X \, \Sigma \, XY}{n \, \Sigma \, X^2 - (\Sigma \, X)^2} \qquad (8\text{-}6)$$

$$a = \frac{n \, \Sigma \, XY - \Sigma \, X \, \Sigma \, Y}{n \, \Sigma \, X^2 - (\Sigma \, X)^2} \qquad (8\text{-}7)$$

If we had been fortunate enough to know that the XY function passed through $(0, 0)$, such that b in Eq. 8-1 was exactly zero, we could have started over and obtained the simpler expression for a alone,

$$a = \frac{\Sigma\, XY}{\Sigma\, X^2} \qquad (8\text{-}8)$$

However, we only show Eq. 8-8 to warn against its use. There is almost no conceivable situation in experimental analysis that justifies forcing the zero on the data. *The intercept of straight-line data is always inherent in the data and should be allowed to express itself.*

Equations 8-1, 8-6, and 8-7 are *regression equations,* and in their application we say that we are *making a regression of Y on X.* If we believe that X has the main precision error, we should simply reverse the axes and plot X as Y and vice-versa.

If we give the symbol Y_{est} to those Y values estimated from Eqs. 8-1, 8-6, and 8-7, with the proper values of X, then we can define a statistic called the *standard error of estimates* S_y,

$$S_{y,x}^2 = \frac{\sum\limits_{1}^{n} (Y - Y_{est})^2}{n - 2} \qquad (8\text{-}9)$$

Equation 8-9 is exactly analogous to Eq. 2-21, defining the standard deviation. The standard error of estimate is simply the standard deviation of all the points in the Y direction above and below the correlation line. We thus expect that $\pm 2S_{y,x}$ in the Y direction will enclose about 95% of all the population of data. This allows us to put a *confidence interval* on a graph of scattered data, something we did in Example 3-1 by replicating tests at two different X values.

One other pair of important statistics in regression studies are the mean values of the X and Y values, $\overline{Y}$ and $\overline{X}$. This point on the XY coordinate system is the *centroid* of the data points, and the regression line passes through it. It is a convenient measure of the center of a group of XY data, especially when comparing several such groups.

If the data are known to follow a logarithmic or other law, the easiest approach in fitting a least-squares line is first to use the methods of Chap. 7 to rectify the data, then fit the new Y and X data to the regression model of Eq. 8-1.

Example 8-2 Anchoring vessels has always been based on the interaction of three principles: weight on the bottom, the accompanying suction that can be generated in soil, and the plow or digging principle. The hydrostatic anchor is a device to obtain the suction without the weight by placing a covered, thin metal, circular box down over the soil and pumping water continuously out of this enclosed space to maintain a lower pressure P_i than that of the surrounding ocean P_{amb}. A screen or porous stone plug prevents soil particles from getting into the pump suction at the top of the anchor. When a dead upward pull is exerted on such a device, the force reaches a maximum F_m and then subsides as the soil fails or the anchor pulls out. Ideally, we would hope that the upward force divided by the active area of the circle F_m/A in newtons per square meter, or pascals, would equal the differential pressure $P_{amb} - P_i$ or ΔP, also in pascals. Unfortunately, the soil usually undergoes shear failure first.

Sixteen tests made on four anchors ranging in diameter from 106.75 mm (4.375 in) to 336.55 mm (13.25 in) and carried out in loose soil with an in-soil skirt equal to half the diameter gave the following data:

F_m/A (Pa)	4,830	3,790	3,790	3,450	4,140	4,140	8,270	8,270
$P_{amb} - P_i$ (Pa)	13,100	14,480	15,860	15,860	16,200	16,550	21,370	22,750
F_m/A (Pa)	6,210	9,310	5,520	8,270	8,960	12,410	11,380	6,550
$P_{amb} - P_i$ (Pa)	23,440	26,200	27,580	29,650	32,410	31,720	35,850	37,230

We wish to fit a least-squares line to these data.

Solution Virtually all the analysis methods of this chapter and the next require either an interactive computer program, such as STATPACK, or an electronic minicomputer with at least eight-place accuracy. Most small memory machines provide a program for regression in their instruction booklets. We might, however, wonder if this data fits our Fig. 8-3 model. If we let Y be F_m/A, we probably put most of the random variation in the Y direction. Finding P_i inside the anchor and P_{amb} outside involves relatively error-free measurements, whereas obtaining the peak force and, more important, keeping the soil exactly the same each time are probably tasks with inescapable precision error. But if we plot the data in Fig. 8-4, we see another problem. The lines marked "probable data envelope" suggest that the precision error is not uniform over the entire range but shows a product-type variation with ΔP. Probably the soil becomes more and more sensitive to discontinuities and small variations as higher and higher ΔP

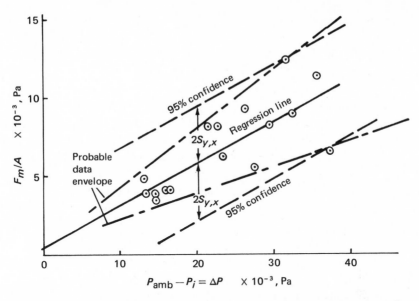

Figure 8-4 Pull-out pressure as a function of pressure difference across the anchor of Example 8-2. The least-squares regression line and the 95% confidence interval (assuming that this is constant, although it clearly is not) are shown.

values are tried. In spite of this, we must fit the regression line. Even when a model does not fit our Fig. 8-3 ideal, the reasons for regression are compelling. It locates a *consistent and reproducible straight line* through the center of the points. It *obtains slope and intercept numbers* that permit the data to be compared with that from other tests or configurations. It *allows various statistical tests to be carried out.* Then we obtain the several quantities for Eqs. 8-6 and 8-7:

$$n = 16$$

$$\Sigma Y = \Sigma \frac{F_m}{A} = 4,830 + 3,790 + 3,790 + \cdots + 6,550 = 109,290$$

$$\Sigma X = \Sigma \Delta P = 13,100 + 14,480 + 15,860 + \cdots + 37,230 = 380,250$$

$$\Sigma X^2 = \Sigma (\Delta P)^2 = (13,100)^2 + (14,480)^2 + \cdots + (37,230)^2$$
$$= 999,846 \times 10^4$$

$$\Sigma XY = \Sigma \left[\frac{F_m}{A} (\Delta P) \right] = (4,830)(13,100) + (3,790)(14,480)$$
$$+ \cdots + (6,550)(37,230) = 285,582 \times 10^4$$

Then, from Eq. 8-6,

$$b = \frac{(999{,}846 \times 10^4)(109{,}290) - (380{,}250)(285{,}582 \times 10^4)}{16(999{,}846 \times 10^4) - (380{,}250)^2} = 438.1 \text{ Pa}$$

And from Eq. 8-7,

$$a = \frac{16(285{,}582 \times 10^4) - (109{,}290)(380{,}250)}{16(999{,}846 \times 10^4) - (380{,}250)^2} = 0.267$$

Thus the "best" straight-line equation for these data is (Eq. 8-1)

$$\frac{F_m}{A} = 0.267(P_{amb} - P_i) + 438.1$$

and is plotted on Fig. 8-4.

The slope a is usually the most important number obtained by a regression, because it gives the proportional relationship between the two variables. Here, we see that for each pascal gain in holding power, about 4 Pa of pressure difference must be added. But as the difference between P_{amb} and P_i grows, water percolation through the soil increases and sets a limit—we may produce pullout by channeling the soil and breaking the suction. Thus tests and design changes are aimed at increasing the slope a, and it is evident that we must obtain this value in as consistent and uniform a manner as possible. Regression using least squares provides this consistent treatment for such data.

We can obtain the standard error of estimate quickly by reading approximate $(F_m/A)_{est}$ on Fig. 8-4. For example,

$Y = F_m/A$ (Pa)	4,830	3,790	3,790	3,450	4,140	$\cdots$
$Y_{est} = (F_m/A)_{est}$ (Pa)	4,000	4,000	4,400	4,400	4,700	$\cdots$
$(Y - Y_{est})^2$ (Pa^2)	688,900	44,100	37,210	902,500	313,600	$\cdots$

Then from Eq. 8-9,

$$S_{y,x}^2 = \frac{688{,}900 + 44{,}100 + 37{,}210 + 902{,}500 + 313{,}600 + \cdots}{16 - 2}$$

and $S_{y,x}$ is 1850 Pa. If we draw lines parallel to the regression line on Fig. 8-4 $(2S_{y,x})$ above and below, we define the zone on the graph between them containing about 95% of the population, that is, the *95% confidence interval*. However, it is easy to see that the lower points will fall closer to the line than this and thus that the upper ones may fall farther out. In any case, there is no justification for rejecting any points, because we do not really have a good idea of the true standard error of estimate and how it varies with ΔP.

Example 8-2 shows how regression can reduce a group of scattered data to one or two significant numbers and thus permit ready comparison with other work. But often the regression equation itself is one of essential importance and is needed to complete the research. An example of a test in which this was the case is shown in Fig. 8-5. Two matched solar cell banks were used to test the effect of augmenting such a bank with a single mirror. It can be shown that if the cell bank is perpendicular to the sun's rays, the ideal concentration ratio (ICR) due to the mirror and defined as the new output divided by the output without the mirror is given by

$$ICR = 1 - \cos\left(2\theta_a\right)$$

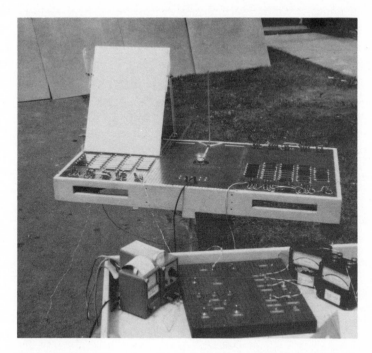

Figure 8-5 Photo of a solar cell concentration experiment. The cell bank on the left receives both direct sunlight and sunlight reflected from the mirror.

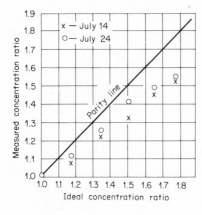

Figure 8-6 A parity graph showing the actual improvement in the cell bank of Fig. 8-5 from the mirror plotted against the theoretical expectation based on geometry only.

where θ_a is the acute angle between the mirror and the plane of the cell bank.[*] If we now run a series of tests on two days in which we measure the concentration ratio by dividing the actual output of the mirror-enhanced bank by the output of the no-mirror bank and plot these against the ideal ratio (ICR) from the noted equation, we obtain Fig. 8-6, another *parity graph*. We see that the actual concentration ratio (CR) is lower than the ICR and, further, that it gets worse as the ICR increases.

Solar cells are known to suffer performance degradation as their temperatures increase. Thermocouples attached to the backs of the cells gave cell temperature during all the taking of data. If we plot the difference between the two concentration ratios against the difference in the temperature between the two banks, we obtain Fig. 8-7. Clearly, we have partly diagnosed the problem, for the two are directly proportioned.

The difficulty was how to correct the actual CR values from some independent data on how cell performance varies with temperature. This was done by comparing each cell bank output with the reading of a *pyrheliometer*, a

[*]Schenck, 1970, pp. 141–162.

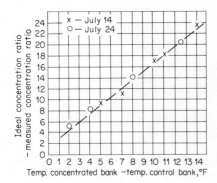

Figure 8-7 Apparent correlation between the error or degraded performance of the mirror bank in Fig. 8-5 with the solar cell temperature.

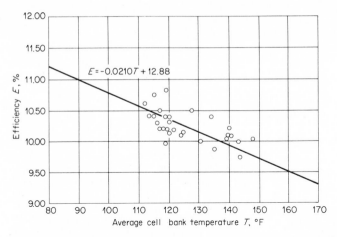

Figure 8-8 Least-squares regression of scattered temperature–efficiency data, where efficiency is obtained by comparing the cell bank with a standard radiation instrument. The least-squares equation is shown.

device calibrated to give the total incident energy reaching a surface from the sun. Then the cell bank output power was divided by this instrument reading to obtain an efficiency E in percent, and the regression line of this E versus cell bank temperature was obtained as shown in Fig. 8-8. There would be no way to correct the actual CR values directly from such scattered data. The regression line organizes it for application. We now use the equation of Fig. 8-8 to obtain the efficiencies of each bank. To correct our measured concentration ratio, we multiply it by the ratio of the predicted efficiency of the augmented bank divided by the predicted efficiency of the no-mirror bank, thereby accounting for the higher temperature of the augmented bank. Figure 8-9 shows that the correction is in the right direction.

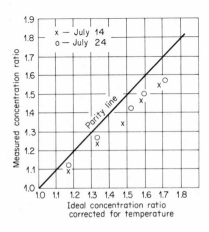

Figure 8-9 Parity graph similar to Fig. 8-6 but now with the ideal concentration ratio corrected by the regression equation shown in Fig. 8-8.

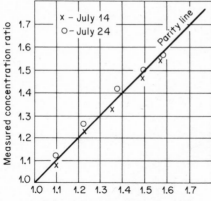

Ideal concentration ratio corrected for temperature, reflection loss at mirror, and reflection loss at solar cell surface

Figure 8-10 A further refinement of Fig. 8-9 in which all losses are included in the ideal concentration ratio.

This Fig. 8-9 correlation can be further improved by accounting for the reflection losses at both the mirror surface and the cell face itself. When this is done, Fig. 8-10 results, and we have discovered and explained the source of all the losses causing the degraded performance in Fig. 8-6, as well as demonstrating the important use of the *parity graph* in the *progressive correction of test data toward a theoretical ideal*. Most essential, however, we see how regression reduces a mass of scattered and unwieldly data to usable equation form.

8-3 CORRELATION

We have already shown how two sets of data can be *correlated*, that is, related on the XY plane by an equation, but we have not really proved that they *are* correlated. We can now imagine a kind of *hierarchy* or *continuum* of precision error or data variability. At the top, we have high-precision data as discussed in Chap. 7. Such data were not only obviously correlated on the XY plane, but we obtained a great variety of equations and constants from it by rectification and function discovery. Section 8-2 dealt with data that have large precision error, but are still quite obviously related. That is, Fig. 8-4 and 8-8 show low-precision data, but there is no real doubt that some sort of functional relationship exists between the anchor breakout force and the pressure difference, or between the solar cell efficiency and the cell temperature.

In this section, we make one further step away from certainty, to data in which the XY relationship is not obvious and to which we must apply further statistical ideas to determine if any correlation exists at all between X and Y.

If we make a regression of Y on X and there is no relationship whatever between them, we shall obtain a horizontal line; that is, the value of the slope a in Eq. 8-7 will be zero. Suppose, now that we make a *regression of X on Y*; that is, we reverse the two variables. The new correlation equation will be

$$X = a'Y + b' \tag{8-10}$$

From the derivation of Eq. 8-7, it is evident that now

$$a' = \frac{n \, \Sigma \, XY - \Sigma \, X \, \Sigma \, Y}{n \, \Sigma \, Y^2 - (\Sigma \, Y)^2} \tag{8-11}$$

That is, we simply reverse the roles of X and Y in Eq. 8-7. How does a' relate to our original assumed line, Eq. 8-1? We can find the answer by solving Eq. 8-10 for Y:

$$Y = \frac{1}{a'} X - \frac{b'}{a'} \tag{8-12}$$

Now if X and Y are not in any way correlated, a is zero; but by the same reasoning, applied to Eq. 8-10, a' must also be zero because changing how we perform the regression cannot change the amount of correlation. But suppose now that we have a perfect correlation, that is, a straight line at some angle on the Y-versus-X coordinates. Now a will be some finite value, and from Eq. 8-12, $1/a'$ will *have to be that same value*, again because changing the order of regression cannot alter the fact of perfect correlation. Thus the product $a \times a'$ is *zero* for *no correlation* and *plus or minus unity* for *perfect correlation*, suggesting that this product is of use in defining the amount or degree of correlation. We can thus define a *correlation coefficient r* by $(a \times a')^{1/2}$ and form it from Eqs. 8-7 and 8-11:

$$r = (a \times a')^{1/2} = \frac{n \, \Sigma \, XY - \Sigma \, X \, \Sigma \, Y}{[n \, \Sigma \, X^2 - (\Sigma \, X)^2]^{1/2} [n \, \Sigma \, Y^2 - (\Sigma \, Y)^2]^{1/2}} \tag{8-13}$$

Of course, almost no data give an r value of zero, because there is always some statistical likelihood of obtaining finite slope values from a sample of data pairs having no true correlation. Statisticians have computed the probability of finding an r value *at or above* given values for various sample sizes (values of n) *if the two populations of X and Y are independent and unrelated*. Figure 8-11 gives curves for probabilities of 0.1, 0.05, 0.01, and 0.001. If, for example, we have 15 pairs of X and Y data and we wish to establish *at the 5% level* that X and Y

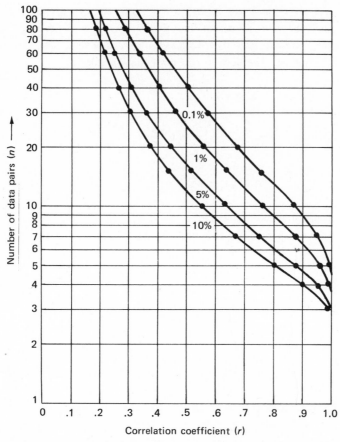

Figure 8-11 A plot showing the value of the correlation coefficient r needed to achieve the given levels of significance with the given number of sample pairs. For example, to suggest that correlation between X and Y exists at the 1% level with 10 pairs of data, r must equal or exceed a value of about 0.76. With 20 pairs of data, an r of only 0.56 would be needed for the same significance level.

have some relationship, we note from Fig. 8-11 that r must exceed about 0.51. There is still 1 chance in 20 that it could exceed this value with X and Y completely unrelated. If we can obtain an r greater than about 0.64, Fig. 8-11 shows that there is now only 1 chance in 100 that we could obtain this with uncorrelated data. This idea of *levels of significance* will be discussed more fully in the next chapter.

Example 8-3 As noted, we assumed that the hydrostatic anchor data of Example 8-2 did show a correlation between maximum breakout pressure and ΔP. Does r for this experiment justify our assumption?

Solution We need one new quantity to apply Eq. 8-13,

$$\Sigma Y^2 = \Sigma \frac{F_m}{A}^2 = 4{,}826^2 + 3{,}792^2 + 3{,}792^2 + \cdots + 6{,}550^2$$

$$= 86{,}583 \times 10^4$$

$$r = \frac{16(285{,}582 \times 10^4) - (109{,}290)(380{,}250)}{[16(999{,}844 \times 10^4) - 380{,}250^2]^{1/2}[16(86{,}583 \times 10^4) - 109{,}290^2]^{1/2}}$$

$$= 0.765$$

Consulting Fig. 8-11 with $n = 16$, we note that there is less than 1 chance in 1000 that an r of 0.765 could come from uncorrelated data, confirming what we can see by eye in Fig. 8-4.

Example 8-4 We treated the two sets of data in Example 8-1 as though they were two separate and unrelated samples. Yet we might expect that those subjects who show high resistance with hand-to-hand measurements would also show high r values with the foot-to-hand test. Is this the case? Can we obtain a regression equation relating foot and hand data, and can we establish that it is correlated?

Solution It is usually a good idea when dealing with random variables like these to plot a *scatter diagram* (X versus Y). Figure 8-12 shows all 15 data so plotted, with the two doubtful points indicated. Using STATPACK, we can rapidly obtain both regression and correlation statistics. Let us do this for two of the cases already considered, that is, first using all the data, and then with the data of subjects 7 and 13 omitted. We shall let the foot data be Y and the hand data be X.

Data used	All (15)	Subjects 7 and 13 rejected (13)
Intercept b, regression of foot on hand	1,823.996 Ω	8,897.383 Ω
Slope a, regression of foot on hand	0.84416	0.33803
Correlation coefficient, r	0.445	0.275
Standard error of estimate, $S_{y,x}$	13,796.938 Ω	7,058.012 Ω
SD of foot, S_y	14,847.164 Ω	7,028.477 Ω

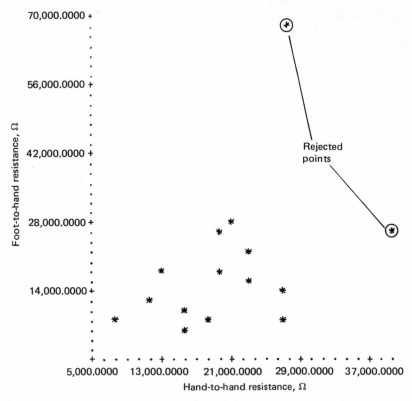

Figure 8-12 Electrical resistance data of Example 8-4 plotted as a scatter diagram.

Notice that it is perfectly possible to get reasonable-looking slopes and intercepts from the data, even though the r values show that they are not correlated. One sure sign of low or no correlation is that the a and b values of the regression equation show huge changes when small amounts of data are dropped. Another sign is that the standard error of estimate is similar in size to the standard deviation of the Y data set as a whole. If there were some relationship between hand and foot data, we would expect that fitting a line and then getting $S_{y,x}$ from it would give a lower dispersion than simply finding the SD from the mean. This is not the case here. Note that with this data set the value of the correlation coefficient r with all 15 data is significant at the 10% level in Fig. 8-11, but when we reject the two doubtful data pairs, we are way below any significance level and can assume that there is no discernible correlation between hand and foot resistance paths in humans. Engineers often assume that data are rejected to enhance or emphasize a relationship. What may just as often occur, as here, is that any suggestion of a relationship or correlation is destroyed by a systematic, statistical, point rejection. In this case, it is mainly the data from subjects 13 and 7 that give a spurious suggestion of direction to the scatter diagram.

This, in fact, is good news to our toy train manufacturer. We have shown that it is not the *internal* path through the subject that is important; if it were, we would expect subjects with high resistance values along one path to have them along another. In fact, we see that it is the foot and hand *contact resistances* that must be important. Heavy people with flat feet will probably show a *low* foot-path resistance but might be too weak to grasp the pipe firmly and thus show a *high* hand-path resistance. Thus our worried toy manufacturer need not assume that someone with unusually low internal resistance may be electrocuted by the trains.

8-4 INSTRUCTIONS FOR USE OF CHAPTER 8

Unless you have run the experiment before or know from other sources what to expect, Chap. 8 is arrived at from Chap. 7. That is, and assuming that the experiment leads to XY-type data and not to pure statistics as in Chap. 9, we plot and decide whether the precision justifies a rectification and function discovery attempt. This is a matter of judgment, but a study of the various data presentations in these two chapters should help in deciding. Assuming then that we can "see" some sort of XY relation, but that we cannot see it as more than a trend, we (1) decide if the data fit our Fig. 8-3 model, or perhaps we want a regression anyway, (2) decide which variable has the most (hopefully all) variability and call this the Y variable, (3) fit the line using the equations of Sec. 8-2, (4) consider the question of statistical rejection of outliers as explained in Sec. 8-1. In the case of outliers from a regression line, we simply use the standard error of estimate $S_{y,x}$, with Eq. 8-9 treating it like the SD of the points around the line. Then, (5) any outlying points that violate Chauvenet's criterion are rejected and the regression run again. At this point, slope and intercept values are obtained and confidence intervals placed on the data using $S_{y,x}$.

When the Y-versus-X scatter diagram shows a cloud of points, it may be necessary to obtain the correlation coefficient (Eq. 8-13) to decide if there is any relation between Y and X at all. Note that *any paired X–Y data can be studied by both regression and correlation*. If correlation is established (see Fig. 8-11), then the regression equations have meaning; that is, the slope of the line is a statistically established indicator of the Y-versus-X trend. If correlation is not established, the regression is a waste of time. This point is demonstrated in Example 8-4.

Since the certainty of correlation is a relative thing, the results of a regression do not always have the same meaning and force. In today's practical experimental efforts, regression often serves as a uniform and reproducible way of obtaining important constants from data.

PROBLEMS

8-1 A device is being tested that follows a law of cooling of the form

$$T = 490e^{-0.077\theta}$$

where T is the temperature in degrees Fahrenheit and θ is the time from the start of cooling in minutes. This law is obtained from the following data using a least-squares analysis:

$\log_e T$	5.5	5.25	4.8	3.5
θ (min)	5	10	15	25

Find the standard deviation of $\log_e T$ from the least-squares predicted line. Decide if any of the data points should be rejected using Chauvenet's criterion. When time was zero, the apparatus was known to be at $570°F$. Does extrapolation, allowing for the scatter expressed by the standard deviation figure, seem to confirm this figure?

8-2 The thermal conductivity of granite is measured a number of times. Of the following 10 readings, 0.23, 0.27, 0.25, 0.23, 0.2, 0.24, 0.31, 0.22, 0.25, 0.21, should one or more be rejected? If so, which one(s)?

8-3 Using least squares, plot the data in Probs. 8-3 through 8-5, and give the straight-line equation.

Machine speed $\times 10^2$ rpm, X	1	2	3	3	4	5
Number of accepted parts, Y	1	3	2	4	4	6

8-4

Pump flow, X (lb/s)	0.5	1.0	1.5	2.0	2.5	3.0
Pump efficiency, Y (%)	1.8	3.2	4.5	5.0	6.4	7.5

8-5

Grid volts, 885 tube, X	0	−5	−10	−15	−20	−25	−30
Plate volts for starting discharge, Y	0	50	100	125	180	250	315

8-6 In Probs. 8-3, 8-4, and/or 8-5, use the line you find through least squares and the deviations from this line to obtain the standard error of estimate $S_{y,x}$. Discuss the intercept(s) in relation to the expectations of the test.

8-7 The average failure time of heavily overloaded resistors was 38.5 min for 52.5-Ω units, 37 min for 53.5-Ω units, 44 min for 54.5-Ω units, and 45.5 min for 55.5-Ω units. Assume that a straight-line function exists between failure time and resistance. Find $S_{y,x}$ of the failure times if all random variation is assumed in their values.

8-8

Flow (lb/s)	3.76	3.38	2.92	2.51	1.64	1.24	0.587
Efficiency (%)	10.5	8.2	7.3	6.5	4.9	3.1	2.1

Decide which points, if any, should be rejected by drawing the best line and assuming that all error is in efficiency.

8-9 Another student lab group of 16 persons measured their body resistance through two paths as in Example 8-1, but on a different day. The following results occurred:

Subject no.	1	2	3	4	5	6	7	8
Hand-hand (Ω)	11,550	10,750	9,000	39,400	55,000	22,500	16,000	19,750
Foot-hand (Ω)	9,700	27,250	9,000	85,400	31,000	15,000	19,000	89,500

Subject no.	9	10	11	12	13	14	15	16
Hand-hand (Ω)	400,000	15,750	45,000	30,800	90,400	15,400	68,570	42,000
Foot-hand (Ω)	7,000	11,800	12,450	9,560	39,570	23,200	16,000	80,000

Obtain the elementary statistics for these two data sets and apply Chauvenet's criterion for point rejection. If any data are rejected (from each column so that correlation can be attempted), recompute the basic statistical material. Without regard to Chap. 9 methods, consider the following questions:

(*a*) Does this set show any significant differences from the Example 8-1 set?

(*b*) Considering the means and the SDs of the means, do you think that this group's means are from the same population as those in Example 8-1?

(*c*) What does this say about the variability of the human body to electric shock?

Speculate on the reasons for the difference between these two groups.

8-10 Using the data in Prob. 8-9, obtain the correlation coefficient of the data, with any points rejected, and show a scatter diagram. Discuss any differences between these results and those of Example 8-1.

8-11 In Example 7-9 an experiment was described which suggested that the optical or turbidity data of water and its suspended-solid constant seemed to be related. A test made in May took water samples and extracted their solid matter by filtration to obtain w in milligrams per liter and also obtained the value of the absorption coefficient α of the water with units of meter^{-1}. This coefficient rises as the water gets dirty and clouded. Within Narragansett Bay the data showed that both quantities increased together.

Solid content w (mg/liter)	8.20	4.6	4.6	9.8	9.68	5.40	8.0	11.4	5.8	10.6	7.4	12.8
Absorption coefficient α (m^{-1})	1.66	1.12	0.76	1.61	1.43	0.82	1.90	2.25	1.57	1.9	1.66	2.40

Obtain the regression line, assuming that the error is concentrated in w, then concentrated in α, and show the lines on a scatter diagram. Does the intercept look reasonable? Are these variables strongly correlated as measured by r? Does this experiment confirm the conclusions of Example 7-9?

8-12 A computer simulation of traffic light operation measures the number of cars in the waiting line every five light changes. Two runs are made, one in which the lights are set to count cars in the lines and change following an instruction program N_p and the other in which the lights are simply operated on a regular timed red–green sequence N_t. Eighty light changes produced the following data:

N_p	16	28	38	34	24	18	18	17	16	19	20	19	18	17	17	18
N_t	16	31	47	45	34	29	25	25	30	35	35	35	37	35	28	22

It is evident that the programmed lights maintained smaller waiting lines than the timed lights. But can we say that this *improvement* increases when there are *more* cars in line? Using the methods of regression and correlation, decide this question.

8-13 Example 8-2 gave data on a hydrostatic anchor in loose soil, relating its pressure difference to the unit area breakout force. Data are now taken on an anchor with a skirt only one-tenth the diameter, as compared to five-tenths in Example 8-2.

F_m/A (Pa)	3,790	5,210	4,820	4,980	6,550	7,240	6,210	6,900
P (Pa)	12,400	13,100	13,700	14,480	13,790	14,130	15,170	15,170

F_m/A (Pa)	7,240	8,270	8,270	8,960	8,620	13,100	10,687	13,790
P (Pa)	15,170	20,680	22,750	21,370	24,130	27,230	26,900	37,920

Decide if these data are sufficiently correlated to justify regression. If so, perform a regression and compare your results, with regard to both anchor performance and the *variation* in anchor performance with the deep-skirted anchor tested in Example 8-2. Should further testing be with deep or shallow skirts in this soil?

8-14 Four new pairs of swimming fins are tested by having five swimmers attempt to swim as hard as possible while attached to a scale, giving maximum thrust T_m, and also to swim against the scale for a period of 5 min at "average" speed, giving T_a. The table gives the parameters of the fins and the results:

Parameter	Fin identification			
	A	B	C	D
Retail cost, Rhode Island	$ 5.49	$11.95	$18.95	$14.95
Area (in^2/fin)	85	61	101	72
Average maximum thrust, T_m (lb$_f$)	16.5	15.9	23.8	20.0
"Average" thrust, T_a (lb$_f$)	11.8	12.0	15.9	14.3

Decide whether there is any correlation between the following variables:

(a) Cost and the two thrusts.
(b) Area and the two thrusts.
(c) T_m and T_a.

If you were the purchasing agent for a retail chain, what would your suggestions be on these items?

8-15 In a study of frost penetration in soil, a guarded hot plate was emplanted at the site to obtain the value of H, the convective heat-transfer coefficient (Btu/h $\cdot$ °F $\cdot$ ft^2) as a function of the site wind velocity V (mph).

V (mph)	0.0	0.0	0.3	0.5	0.7	1.0	1.0	1.0	1.5	1.9	2.1
H (Btu/h $\cdot$ °F $\cdot$ ft^2)	1.9	2.05	2.5	2.5	2.5	2.5	2.7	2.2	2.1	2.4	2.5

V (mph)	3.1	3.1	4.2	4.3	5.2	5.25	6.3	6.6	6.6	9.8	9.8
H (Btu/h $\cdot$ °F $\cdot$ ft^2)	2.9	2.4	2.95	2.7	2.7	2.5	2.7	2.8	2.5	2.7	2.9

Are these data related significantly? We need a regression equation to estimate H on the site between 0 and 10 mph. What will be the 95% confidence interval if H has all the error?

8-16 In a study of wave-induced surface drift, a series of thin plastic sheets of several dimensions were observed to move with a velocity V with waves of 1-in height and 1.3-ft length.

Drift rate (ft/s)	Object dimensions (in)				
	(3 × 4)	(3 × 6)	(3 × 8)	(3 × 10)	(3 × 11.75)
2 April	0.110	0.111	0.133	0.150	0.147
3 April	0.097	0.105	0.125	0.153	—

Obtain the regression lines for these data, assuming that all variation occurs in drift rate. Is object length or area a better parameter? Are the laws obtained from the two days different, or should the nine data be treated as a group?

8-17 A solar module is tested under actual sunlight, and the incident solar radiation is obtained in joules per second per square meter by a pyrheliometer. The short-circuit cell current is expected to follow a straight-line relationship with input.

Solar input ($J/s \cdot m^2$)	489	552	631	656	662	668	684	703	709	725	694	757	779
Short-circuit current (mA)	27	32	36	37	36	37	38	39	42	42	46	39	46

Perform a regression on these data, assuming that the output (current) has all the variability. The data must give zero output for zero input. Is this intercept justified by the data? Plot the data and the regression line and add the 95% confidence limits on the data.

8-18 In an experiment involving two persons breathing in a restricted space, the carbon dioxide concentration in percent is measured at various times after the start of the experiment:

Percent CO_2	0.13	0.13	0.13	0.18	0.22	0.20	0.25	0.32	0.45	0.54	0.37	0.54
Time (min)	2	10	15	18	20	25	28	36	40	48	70	76

Obtain the regression line assuming that the error is concentrated in percent CO_2. If discomfort occurs at 4.0% CO_2, predict how long it would take at these conditions to reach that point. With regard to the intercept, does the data suggest that we began with no CO_2 or with some amount already present?

8-19 A home heating duct is tested for loss in temperature versus different lengths of run. The following data result for a given type of duct and fixed surrounding conditions:

Temperature loss (°C)	5	7	15	20	22
Length (m)	4	8	12	16	20

Find the equation, assuming a straight-line function with all variation in temperature loss. Show the plot and indicate the 50% confidence interval on it.

BIBLIOGRAPHY

Acton, F.: *Analysis of Straight-Line Data,* Wiley, New York, 1959.

Bacon, R.: "Best" Straight Line among the Points, *Am. J. Phys.,* vol. 21, no. 6, pp. 428–445, September, 1953.

Bennett, C. A., and N. L. Franklin: *Statistical Analysis in Chemistry and the Chemical Industry,* chap. 11, Wiley, New York, 1954.

Bowker, A., and G. Lieberman: *Engineering Statistics,* chap. 9, Prentice-Hall, Englewood Cliffs, N.J., 1959.

Deming, W. E.: *Statistical Adjustment of Data,* chap. 9, Wiley, New York, 1943.

Dumond, J. W.: Review of *An Introduction to Scientific Research, Am. J. Phys.,* vol. 21, no. 5, pp. 393–394, May, 1953.

International Business Machines: "Call/360, Statistical Package (Statpack) Version 2," International Business Machines Corp., White Plains, N.Y., 1970.

Proschan, F.: Rejection of Outlying Observations, *Am. J. Phys.,* vol. 21, no. 7, pp. 520–525, October, 1953.

Schenck, H.: *Case Studies in Experimental Engineering,* McGraw-Hill, New York, 1970.

Spiegel, M. R.: *Probability and Statistics,* Schaum's Outline Series, McGraw-Hill, New York 1975.

Wilson, E. B.: *An Introduction to Scientific Research,* chap. 8, McGraw-Hill, New York, 1952.

Worthing, A. G., and J. Geffner: *Treatment of Experimental Data,* Wiley, New York, 1943.

NINE

STATISTICAL DATA ANALYSIS

We arrive in the last chapter of this book at the same topic with which we began Chap. 2, the statistical behavior of a random variable. We saw in Chap. 7 that high-precision data give us large amounts of information about the meaning of a test, specifically, its *function*, its *constants,* and its *projections.* Then in Chap. 8 we saw that there is a class of lower precision data that forces us to choose a very simple function or to be unable to detect the type of function at all. We were, however, still able to get constants of a sort, the slope and intercept, but we would certainly never attempt much in the way of projections, especially to infinity. One further step down the ladder of increasing precision error was noted in Sec. 8-3. There we could say nothing more about the data than that they were, or were not, related.

With this chapter we arrive at the bottom rung of the ladder, data about which no graphical, that is, *X*-versus-*Y*, sense remains. We saw in Examples 8-1 and 8-4 how data on the electrical resistance of persons by two different paths, not only did not give a useful *XY* plot, but was not correlated at all. There are still questions to be asked about such data. For example, are the two paths *different* in electrical conductivity? Are the variations of the data, the SD values, different? These are pure statistical questions, and their answers are usually given by probability numbers, that is, by statements about the expectation of something being the same or different. We did this in Chap. 8 when we spoke of the *chance* of a correlation coefficient being above a certain value.

Such data, low in precision though they may be, encompass a vast array of important studies and enterprises. Questions about the safety of drugs, additives, and nuclear power, about the choice of materials, methods, and costs in making most products, about the best way to structure ambulance service or crop rotation, all derive answers from studies of random data in which we seek a difference or the lack of a difference.

9-1 STATISTICAL TERMINOLOGY:
TWO KINDS OF INFERENCE ERROR

In Chap. 2 we discussed such important ideas and terms as the population and the sample, the normal law, deviations, the mean value, and so on, most of these being statistical conceptions. We shall now use statistical ideas, not as ways of finding and specifying precision error in instrument systems, but as full-fledged means of analyzing the entire test, and some additional terminology must be considered. Most important is the idea of the *significance* of a test or test result. In engineering terms, we might say that a test showing that 20 samples of steel A failed at 60,000 ± 5,000 psi whereas 20 samples of steel B failed at 80,000 ± 5,000 psi was highly significant in that it proves steel B to be stronger. Statisticians would not disagree in this case, although they might suggest *tests of significance* that would allow us to represent this highly significant result by a number. Should the 20 samples of steel B show a strength of 63,000 ± 5,000 psi, we might begin to wonder if we have indeed proved steel B to be stronger, and a statistical approach becomes more necessary. Once we make such statistical tests, and several will be described in the following sections, we admit the possibility of two kinds of inference errors. Errors of the first kind, or *type 1 errors,* result if we ascribe a real effect to differences where no real effect, in fact, is present. Thus, we might say that sample B (with strength 63,000 ± 5,000 psi) is definitely stronger than sample A and proceed to buy only type B steel and use it. Later work (larger test samples) might show that there is actually no real difference between the steels, so that our choice of B on the basis of the original 20-item sample was an error of the first kind. We can reduce the chance of type 1 errors by demanding that our test results be very significant, but we then increase the risk of errors of the second kind, or *type 2 errors,* which occur if we miss a real effect or difference that does, in fact, truly exist. We might, for example, decide that the difference between steel lots A and B is not significant and use the two indiscriminately. Later tests might then show that steel B is unquestionably stronger than steel A in which case we have made a type 2 error.

These ideas may seem rather obvious, but they bring out an important aspect of statistical inference, especially in engineering situations. Most engineering testing will lead eventually to action, and there is no hard-and-fast rule we can make regarding the relative undesirability of type 1 versus type 2 errors. If we are running tests on parachute-opening devices, basic humanitarian motivation should force us to reject the possibility of type 2 errors (which would be the failure to select the truly best device), whereas a type 1 error (selecting the most costly device, which seemed to, but did not, behave more dependably than the others) is not so serious. Conversely, when we are choosing among several kinds of electrical relays to place in an inexpensive child's toy, a type 2 error (failure to select the best relay) is probably not so serious as a type 1 error (choosing a specific relay design that is not really better than the others but is more expensive).

9-2 THE χ^2 TEST FOR SIGNIFICANCE

There are many test situations from which we obtain data in the form of *numbers of items.* Some numbers of parts pass or fail inspection, complete or fail to finish life tests, sell or are returned unsold. Inspection methods may miss some number of reject parts and find some other number, or they may reject some good parts and pass some other number of good parts. Shifts, personnel, machines, assembly lines, production methods, and so on, may produce different numbers of items over a given period. Cosmic-ray counts, numbers of tests, vehicles, or parapsychologically guessed card faces may vary from person to person, day to day, shift to shift, location to location, and so on. Many of these kinds of tests can be analyzed for significance using the *chi-square (χ^2) test.*

The derivation of the χ^2 distribution (see Fig. 9-1) will not be given here but can be found in most complete statistics textbooks.* To apply this distribution, we first compute χ^2 from the equation

$$\chi^2 = \Sigma \frac{(O - E)^2}{E} \tag{9-1}$$

where O represents the *observed* number of occurrences (failures, sales, rejects, items made, counts, correct guesses, etc.) and E represents the *expected* number of these same occurrences. Now, the minute we talk about an expected number, we introduce the idea of the *hypothesis* that we are trying to prove true or false. Suppose, for example, that we wish to find out whether an old or a new lathe will give more good parts. We make the same number of parts on each lathe and then our hypothesis to be tested might be: "Both lathes turn out the same number of good parts." Or, from our previous experience we might choose: "The new lathe turns out one-third more good parts than the old one." Either

*For example, Mood and Graybill, 1963.

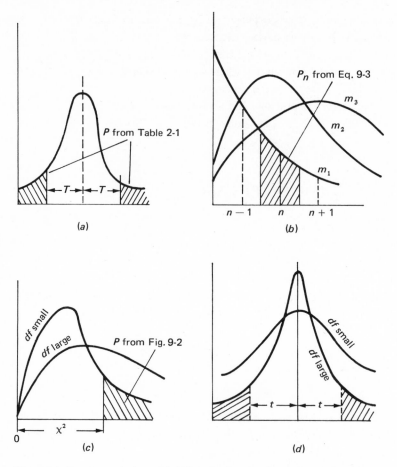

Figure 9-1 Sketches of the probability density functions for (*a*) normal, (*b*) Poisson, (*c*) χ^2 distributions, and (*d*) Student's *t* showing the effect of the number of degrees of freedom.

hypothesis could be tested by an appropriate χ^2 computation. Before seeing how this is done, however, we must consider one additional new term. Any tabular or graphical representation of the χ^2 distribution requires a knowledge of the number of *degrees of freedom* (*df*) present in the test. The degrees of freedom are the number of *independent* groups of observations that can be hypothesized. This is always puzzling to beginners, and it is worthwhile here to give some examples.

A buyer has fractional-horsepower motors installed, half from company A and half from company B. After some period of time, F_a failures have occurred in the company A motors and F_b have occurred in the company B motors. The total number of motors in the failure sample is fixed and equal to $F_a + F_b$. Suppose that we wish to test the hypothesis that motor failure rates are the same. Then our expected F_a can be chosen as $(F_a + F_b)/2$, and the expected F_b must take the value $(F_b + F_a)/2$. Because we are able to choose only one number, we have one degree of freedom and χ^2 becomes[*]

$$\chi^2 = \frac{[F_a - (F_a + F_b)/2]^2}{(F_a + F_b)/2} + \frac{[F_b - (F_a + F_b)/2]^2}{(F_a + F_b)/2} \qquad (9\text{-}2)$$

A test is operated around the clock. Shift 1 takes half of all the points and makes N_1 reading errors; shift 2 takes one-third of all the points and makes N_2 reading errors; shift 3 takes one-sixth of all the points and makes N_3 reading errors. There are N errors made in all, and we wish to test the hypothesis that there is no difference among the shifts in making errors. If such a hypothesis is correct, the chosen expected number of errors for shift 1 is $N/2$, because taking half the points should result in shift 1 making half of all the errors. We choose for shift 2 an expected error number of $N/3$, and this leaves shift 3, $N/6$ errors, because these are what is left of the N total. Thus 2 df fit here and χ^2 becomes

$$\chi^2 = \frac{(N_1 - N/2)^2}{N/2} + \frac{(N_2 - N/3)^2}{N/3} + \frac{(N_3 - N/6)^2}{N/6}$$

Sometimes we have data on two populations or sets of occurrences in which some factor is changed between them, and we are interested in whether the failures or rejects are significantly different in the two cases. Thus, samples of a steel lot are case-hardened (A many) or left alone (B many). These data are most easily handled in a *contingency table* as follows:

Observed data

	Passed	Failed	Total
Case-hardened	$A - X$	X	A
Left alone	$B - Y$	Y	B
Total	$A + B - X - Y$	$X + Y$	$A + B$

[*]When 1-df situations occur, the practice is to subtract 0.5 from each $O - E$ term before squaring. Failure to do this results in a slightly high χ^2 value but produces small error in most engineering computations. Since we are seldom sure just how "significant" we want to be, this slight correction will not be included in the examples that follow.

Now, if our hypothesis is "Case hardening has no effect," we draw up another table, which reflects the expected values of each of the entries or *cells* in the table.

Expected results

	Passed	Failed	Total
Case-hardened	$A - \dfrac{A}{A + B}(X + Y)$	$\dfrac{A}{A + B}(X + Y)$	A
Left alone	$B - \dfrac{B}{A + B}(X + Y)$	$\dfrac{B}{A + B}(X + Y)$	B
Total	$A + B - X - Y$	$X + Y$	$A + B$

Notice that in filling our expected-result table, we are restricted by the need to keep the four totals the same as in the observed-data table. This restriction, examination will reveal, allows us to choose only *one* of the four cell values, because once we choose one, the fixed totals force the values of the other three upon us. Thus *one degree of freedom* holds for this case, and χ^2 is summed from the four terms derived from the expected and observed cell values and Eq. 9-1.

Once we have obtained these two crucial quantities, the number of degrees of freedom, and the value of χ^2, we enter an appropriate table[*] or Fig. 9-2 and read out the *probability that the χ^2 value can be equal to or greater than the one found, owing to chance*. If this probability figure is, say, 10, 20, or 30%, this means that the hypothesis is reasonable (*not* proved, however). In any event, such a probability value suggests that our experimental values and the values based on our hypothesis are *not* from different populations. A probability figure of 5% or less, on the other hand, begins to cast doubt on the hypothesis. The 5% figure, for example, suggests that our experimental distribution, or one that is even more distorted, could not occur by chance from our hypothesized distribution, data, or sample more than 1 in 20 times. For a *significance level* of 1%, the chance of occurrence shrinks to 1 in 100 times, and so on. The level of significance that is appropriate depends on the test, its use, and the number of data available. Speaking generally, 5% raises doubts, and 1% approaches certainty. Notice in this discussion that we are most often attempting to disprove the hypothesis.

Suppose that, in the first of the three examples (dealing with motors), we

[*]For example, Brownlee, 1953.

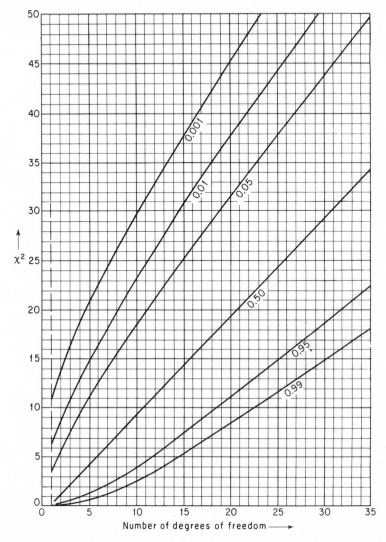

Figure 9-2 A plot showing χ^2 as a function of the number of degrees of freedom and of the probability of occurrence of a given χ^2 value. This table is based on the chance of finding the χ^2 value in the tail, as Fig. 9-1 shows.

note five failures from company A motors and nine from company B motors. Then Eq. 9-2 gives

$$\chi^2 = \frac{(5-7)^2}{7} + \frac{(9-7)^2}{7} = 1.14$$

Entering Fig. 9-2 with this value and 1 *df* (*n* of 1), we find a probability definitely greater than 10%, and our hypothesis is not disproved. Suppose now that more time passes, and the same ratio of A-to-B failures continues until we note 15 A failures and 27 B failures. Now χ^2 is 3.43, the degrees of freedom are unchanged, and the probability of this value (or of a greater one) has dropped to almost 5%. Although this still does not definitely disprove our hypothesis, our suspicions should be aroused. Further operation and failures at this same rate will suggest with increasing assurance that our hypothesis is false and will allow us to say what we felt all along, namely, that company A's motors are better.* Thus the χ^2 test is quite sensitive to the size of the sample, and often more data are needed before real significance is attained.

One other precaution might be mentioned here. *The minimum size of the expected frequency in a cell is five units.* Thus, if $A + B$ motor failures had not reached the number 10, we could not properly have used the χ^2 test.

The χ^2 distribution starts at zero and is single peaked with a long "tail," the extent of this tail depending on the number of degrees of freedom. Figure 9-1 shows a sketch of the density function. Figure 9-2 gives the probability of finding a χ^2 value out in the tail, and the farther we go, the less chance there is of such a value and the less chance that the O and E data are from the same population. Some statistics books suggest that if the χ^2 value is too low, so that we find a very high probability, say 0.99, we should view the data with suspicion. If, for example, we threw dice and then compared their output with that expected by chance, a very high probability would suggest that the data were "suspiciously" good. It is difficult to imagine a laboratory-type experiment that would be rejected because the data were "too good." If one uses the χ^2 test for comparing random data to the Poisson distribution, it is quite easy occasionally to get data that compare at the 95% (or better) level of confidence. Example 9-2 is a case of this sort. All that this means is that four or five O numbers and four or five E numbers in the four or five cells of the χ^2 expression happened to be very close to each other.

Some tables and graphs of the χ^2 distribution are the reverse of Fig. 9-2; that is, they give the probability of finding a χ^2 value between zero and the χ^2 value rather than out in the tail. To convert, we simply subtract the Fig. 9-2 values from 1.0. Thus the 0.05, or 5%, line becomes a 0.95, or 95%, line.

---

*It is assumed here that motor test is done *with replacement*; that is, when a motor fails, it is immediately repaired and put back in use so that the half-and-half division is maintained. If a failed motor is not replaced or repaired, the company A motors will gradually become the majority, and their failure rate, compared with company B motors, will increase. This case is too subtle for elementary treatment.

Example 9-1 In Sec. 2-10 we discussed an experiment involving the location of a sound source underwater by a diver. Figures 2-13 and 2-14 showed that the complete distribution of points did not pass the test for normality, thus suggesting some interpretations of the experiment. These data can, however, be broken down or grouped, because the main point of this research was, not only to detect acoustic directionality underwater, but to study its enhancement. If we note that the *probable error* of the distribution in Fig. 2-13 is about 40°, we can assign one of two outcomes to each diver point trial. If the diver points closer than 40°, this is a *success*, that is, better than chance expectation; and if the diver points farther away, a *failure*. The most significant-looking preliminary data involved the effect of sound frequency. If, for example, we segregate all 38 data made at 800 Hz, we find 28 successes and 10 failures, but if we pool the data at 2000 and 4000 Hz, the highest practical frequencies used in the experiment, we obtain 12 successes and 26 failures. This difference looks very significant. Can we assign a probability value to the hypothesis that there is no true difference between high and low frequencies with regard to acoustic direction sensing in the water?

Solution This is exactly the same type of data already described in the discussion of the *contingency table*. Thus,

Observed data

Treatment	Success	Failure	Total
800 Hz (low)	28	10	38
2000 + 4000 Hz (high)	12	26	38
Total	40	36	76

Our hypothesis is that the treatments make no difference in pointing success, thus:

Expected results

Treatment	Success	Failure	Total
800 Hz (low)	$38 - [(38/76)36] = 20$	$(38/76)36 = 18$	38
2000 + 4000 Hz (high)	$38 - [(38/76)36] = 20$	$(38/76)36 = 18$	38
Total	40		36	76

Then we form χ^2 from Eq. 9-1:

$$\chi^2 = \frac{(28 - 20)^2}{20} + \frac{(10 - 18)^2}{18} + \frac{(12 - 20)^2}{20} + \frac{(26 - 18)^2}{18} = 13.52$$

With 1 df, as is always the case for a 2×2 contingency table, we see that the probability in Fig. 9-2 for this χ^2 value is much less than 0.001. We can thus say with considerable assurance that low frequencies do lead to lower pointing errors.

Based on this result, the experiment was continued at 800 Hz with five additional treatments: (1) pulsed wave, $\frac{1}{2}$ s on, $\frac{1}{2}$ s off; (2) pulsed wave, $\frac{1}{8}$ s on, $\frac{1}{8}$ s off; (3) pulsed wave, 1 s on, $\frac{1}{4}$ s off; (4) continuous wave with rubber hood having ear holes cut out on diver; and (5) pulsed 1 s on, $\frac{1}{4}$ off and with hood.

Observed data

Treatment (all 800 Hz)	Success (S)	Failure (F)	S/F	n
Continuous tone	28	10	2.8	38
(1) $\frac{1}{2}$ s on, $\frac{1}{2}$ s off	20	16	1.25	36
(2) $\frac{1}{8}$ s on, $\frac{1}{8}$ s off	22	14	1.57	36
(3) 1 s on, $\frac{1}{4}$ s off	47	14	3.36	61
(4) Hood	25	11	2.27	36
(5) Hood + 1 s on, $\frac{1}{4}$ s off	12	0	–	12

It is evident that treatment 3 gave the highest success ratio and it is reasonable to ask if this is significantly better. Comparing cols. 1 and 3, we obtain for the expected results.

Expected results

Treatment (all 800 Hz)	Success	Failure	Total
Continuous tone	28.8	9.2	38
1 s on, $\frac{1}{4}$ s off	46.2	14.8	61
Total	75	24	99

Obtaining the observed values from the table of treatments, we compute χ^2,

$$\chi^2 = \frac{(28 - 28.8)^2}{28.8} + \frac{(10 - 9.2)^2}{9.2} + \frac{(47 - 46.2)^2}{46.2} + \frac{(14 - 14.8)^2}{14.8} = 0.149$$

This unreadably low value on Fig. 9-2 simply means that the success–failure outcomes are very similar and that the apparent improvement in the S/F ratio of from 2.8 to 3.36 has no real importance. In fact, the only pair of data within the treatments table (leaving out treatment 5 for the moment) that is unlikely to have come from the same population at the 5% level is treatments 1 and 3 ($\chi^2 = 4.19$, $0.01 < P < 0.05$). Of course, all that this means is that we can fish through the data and selectively "prove" something. The important finding is that the pulsed tones are not significantly worse or significantly better than the continuous tones, and the hooded-subject data are not in any way significantly different from the basic or reference data.

We might expect that at least the final treatment, which shows a perfect success record, would show a significant difference. If we compare treatment 6 with the continuous-tone data, we obtain for the expected-results table:

Expected results

Treatment (all 800 Hz)	Success	Failure	Total
Continuous tone	30.4	7.6	38
Hood + 1 s on, $\frac{1}{4}$ s off	9.6	2.4	12
Total	40	10	50

Then, using the observed data in the treatment table, we obtain a χ^2 of 3.97, just barely outside the 5% line on Fig. 9-2. Thus this apparently perfect score is only marginally significant, and if we look at the other, more extensive data, we suspect that a few more trials would have probably added to the zero and reduced the χ^2.

What we have shown, then, within the limitations of the sample sizes, is that frequency has a large and measureable effect on pointing direction under water, but that no other attempted treatment appears to enhance or diminish the pointing error. As we shall see, this statistical conclusion can be arrived at in other ways, but this method of using the probable error to divide the data is one of the easiest.

9-3 THE POISSON DISTRIBUTION

In Sec. 2-10 we saw how data from a complete test could be compared with the expectation of the *normal distribution*, using probability coordinates. We could, after finding the mean and the SD of the sample, use Table 2-2 to predict what fraction of the 350 data points would be expected at various angular deviations and then compare these expected frequencies with those actually observed, using the χ^2 statistic.

There may be cases in manufacturing or experimental work where certain occurrences happen by chance, now in profusion, now not at all. Cosmic-ray counts, bacterial densities, traffic accidents, and equipment malfunctions are examples of such occurrences. On the other hand, there may be occurrences that seem to have a random character but that actually are following some deterministic scheme or pattern. The number of static bursts per hour on a radio monitor could be an entirely random matter, but it might be affected by rush-hour traffic or by local television-viewing habits. The failure rate per day of pay telephones in Grand Central Station might be a truly random distribution, but it could perhaps be a function of commuter-train schedules or even police surveillance patterns. When a set of occurrences or counts is suspected as being a purely random group, a common test for such a hypothesis is to compare the experimental distribution with an appropriate form of the *Poisson distribution,* a nonnormal distribution derived either from assumptions of chance effects[*] or from yet another distribution known as the binomial, which we shall not consider. Should the experimental distribution be very close in form to the expected Poisson type, we suspect that occasional large numbers in the distribution are not freaks but simply samples from a proper Poisson-type population. Such a conclusion thus saves us from examining the periods or tests in which such large numbers occur in hope that we may find some special and intermittent effect producing them. The usual normal distribution is inappropriate here, because often we are dealing with periods of time or regions on microscope slides, maps, or graphs, in which there may be zero counts or occurrences, and with situations in which zero is a lower bound to the distribution. The normal distribution is symmetrical about its central point with tails extending to plus-or-minus infinity. Thus the Poisson distribution is a skewed but still a chance distribution (see Fig. 9-1).

If P is the total number of occurrences, malfunctions, or accidents, and N is the total number of time periods or regions considered (some of which may have zero occurrences or counts), let the average occurrence (or count) per time period (or region) be m and be equal to P/N. Then the following table will give the expected probability of occurrence per period based on the Poisson distribution:

[*]See Cramer, 1955, pp. 102–108.

Number of incidents or counts in one time period, region, etc.	Probability of this number of time periods
0	e^{-m}
1	$me^{-m}/1!$
2	$m^2 e^{-m}/2!$
3	$m^3 e^{-m}/3!$
n	$m^n e^{-m}/n!$

This set of terms forms a Poisson series summing to unity. Thus, each term represents the probability that the given number of occurrences, counts, or malfunctions will occur. When m is less than unity, the Poisson distribution has a maximum probability at zero occurrence or count. It peaks at one occurrence with m between 1 and 2, and so on. The Poisson distribution is usually checked for by finding m, computing each cell value, and comparing these term by term with the observed distribution, using a χ^2 test. For rough engineering purposes, we might prefer a graphical test similar to the check for normality using probability paper.

Ignoring the zero-occurrence term for the moment, we see that the actual number of expected time periods or regions for the nth term E_n is (with N, as noted, the total number of such periods)

$$E_n = \frac{Nm^n}{e^m n!} \tag{9-3}$$

This expression can be transformed to

$$\log E_n = n \log m + \log N - m - \log n! \tag{9-4}$$

or

$$\log (E_n \times n!) = C_1 n + C_2$$

where C_1 is a constant equal to $\log m$ and C_2 is a constant equal to $\log(N-m)$, with neither being a function of n, the term number. Equation 9-4 is then the familiar slope-intercept equation of a straight line. Thus, if we wish to check whether a given distribution is Poisson, we do not need to compute m. Taking semilogarithmic coordinates, we multiply the observed occurrence number E_n at each factorial and plot this number on the log scale against n on the linear scale. (For the $n = 0$ time periods we let $0!$ equal 1.) Then if a plot of $E_n \times n!$ versus n is a straight line, or a reasonable approximation thereof, we suspect that we are dealing with a Poisson series. And we further suspect that the incidents or counts are the result of purely random or chance factors. Two examples will give this kind of statistical inference some objective reality.

Example 9-2 We are given the following data on cosmic-ray counts from coincident Geiger counters:

Counts in 1 min, n	0	1	2	3	4	5 or more
Numbers of minutes[a] having n counts, E_n	13	22	14	7	4	0

[a]Total = 60 intervals.

Cosmic-ray data in this basic form are expected to be completely random and therefore to follow a Poisson series. Can we say that this is true of these data?

Solution Let us first attempt our graphical test, as follows:

n	$n!$	No. of minutes, E_n	$E_n \times n!$
0	1	13	13
1	1	22	22
2	2	14	28
3	6	7	42
4	24	4	96
5	120	0	0

Now, a plot of the first column n on linear coordinates against the last column $E_n \times n!$ on logarithmic coordinates constitutes our test and is shown in Fig. 9-3. The first four points form a very good straight line. The fifth point is high, but notice that a single count here will make a very large difference. Thus the low-n points are the most significant and indicate that a good Poisson distribution has resulted. A more rigorous test is the use of χ^2. The average number of counts m is $87/60$, or 1.45. Using this, we can compute a predicted value for each cell in the table from Eq. 9-3, obtaining

n	E_n, observed	E_n, Poisson prediction
0	13	14
1	22	20
2	14	15
3	7	7
4	4	3
5	0	1
Total	60	60

where the predicted values have been rounded to the nearest whole number. Grouping the last three cells to obtain an E_n value > 5.0,

$$\chi^2 = \frac{(13-14)^2}{14} + \frac{(22-20)^2}{20} + \frac{(14-15)^2}{15} + \frac{(11-11)^2}{11} = 0.215$$

with 3 df. From Fig. 9-2 we see that the probability of occurrence of this χ^2 value or one higher is almost 0.99. Thus our hypothesis that these two distributions are from the same parent population is almost a certainty, and our χ^2 statistic agrees with our rough graphical test.[*]

[*]Data from Guest and Simmons, 1953. The writers also show how a similar analysis can be applied to the distribution of time intervals between successive particles.

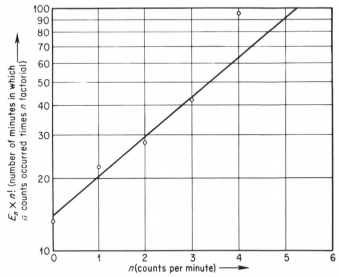

Figure 9-3 A Poisson plot of cosmic-ray data from Example 9-2. The straight line indicates the predicted point distribution for a perfect Poisson series having the value of m in the sample.

Example 9-3 A total of 647 women employed in a factory are observed over a given period. The number of accidents that each suffers is tabulated and the following distribution noted:

No. of accidents, n	Women suffering n accidents, E_n
0	447
1	132
2	42
3	21
4	3
5	2
6 or more	0
Total	647

We expect accidental or chance happenings to follow a Poisson distribution. Is this true here?

Solution For our graphical test, we tabulate as before:

n	$n!$	E_n	$E_n \times n!$
0	1	447	447
1	1	132	132
2	2	42	84
3	6	21	126
4	24	3	72
5	120	2	240
6 or more	720	0	0

Plotting again the first versus the last columns on the appropriate co-ordinates, we see little resemblance to a straight line, even though the tabulated data look "Poisson-like" to the casual eye. A χ^2 test can be made as before, and if this is done, it will be found that the probability that this distribution came by chance from a Poisson series is much less than 0.001. Thus we conclude that the accident rate of these factory workers does not follow a Poisson distribution. The reason is interesting and may give some insight into one reason why certain kinds of distributions are not of the simple Poisson type. If each of the 647 women is exposed to about the same accident risk and, furthermore, if each has the same propensity for accidents, the chance that any given woman might have an accident is identical with that of any other woman. If having one accident does not affect the woman's later behavior very much, her chance of having a second accident when she goes back to work is about the same as any other person having a first, and so on. Thus some persons will be "unlucky" and have one, some will be even less lucky and have two, and so on. But Fig. 9-4 shows that there are too many women with high numbers of accidents. We might not be surprised if these points fell below the Poisson line, for this would suggest that having one accident would lead to greater caution and mitigate against a second.

What we have found here, of course, is the phenomenon of the "accident-prone" personality. Our 647 women do not fall into a homogeneous single group in accident propensity. There are probably several groups present having varying accident propensities. Thus the basic assumption that any woman is as likely to have an accident as any other is demonstrably untrue. Many women who have three, four, or five accidents quite obviously belong in a separate population* or populations.

*Data from Greenwood and Yule, 1920. This is actually a *compound Poisson distribution* and a close fit can be achieved by assuming that the women fall into two separate groups with regard to accident propensity. This distribution is a matter for advanced statistics.

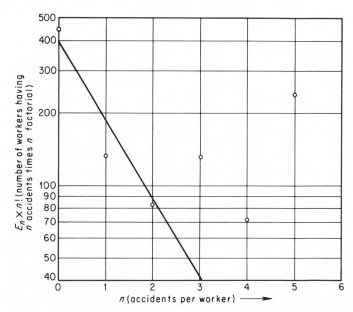

Figure 9-4 A Poisson plot of the accident data from Example 9-3. The straight line indicates the predicted point distribution for a perfect Poisson series having a value of m in the sample.

9-4 STUDENT'S t TEST

The χ^2 test allowed us to compare observed-frequency data to various ideal models, ones that assumed a perfectly random distribution or else a defined statistical distribution. There are many distributions of data that are continuous in nature, such as our data in Example 8-1 on bodily resistance by two different paths, and we might wish to consider if the means of these two are really different. One way to look at such comparisons was shown in Sec. 2-7 and Eq. 2-25 giving the standard deviation of the mean. We noted that there were 2 chances in 3 that the population mean lay within 1 SD on either side of the sample mean. *Student's* t statistic permits direct comparison of two distributions or of a distribution and a fixed population mean value. The t distribution looks very much like the normal distribution, having symmetric tails and a central peak (Fig. 9-1). It was created for use with small (less than 30 items) samples to give better estimates than the normal distribution.

Because the t distribution has two tails, we have the possibility of a *one- or a two-tailed test,* just as was possible with the normal distribution. There we were interested in the probability that a reading might lie *only* above or *only* below a certain deviation T or that the reading might lie either above or below the deviation. Suppose that we asked what positive deviation T would shut out 5% of the deviation. As shown in Fig. 9-1, we would look in Table 2-2 and find the number of SDs, the T value, above and below the mean. If, however, we were asked what T value shut out 5% of the distribution in the *upper* tail, we would not need as large a T, because this is equivalent to shutting out 10% from the two tails. The same is true of the t distribution. We may ask whether two means are simply *different, a two-tailed test,* or whether one mean is *larger (or smaller) than the other, but not both, a one-tailed test.* Obviously, if we demand that our t statistic lie outside the $\pm t$ value that excludes 5% of the distribution, that t value will be larger than if we demand that it exclude 5% of the distribution in, say, the upper tail only. The one-tailed test is thus a less stringent test than the two-tailed one, allowing twice the probability for "significance."

The t statistic can be used for several kinds of hypotheses, but we shall consider only two. If we have a known mean of a *population* μ_X and wish to compare the mean of a sample $\bar{X}$ to it, we find t from

$$t = \frac{\bar{X} - \mu_X}{s_x} \, n^{1/2} \qquad (9\text{-}5)$$

where s_x is the SD of the n items making up the mean $\bar{X}$, and the number of degrees of freedom is simply the number of items n less one, or

$$df = n - 1 \qquad\qquad (9\text{-}6)$$

An example of this use of the t test involves the data of Sec. 2-10 where we expected that the mean of 350 pointing trials by scuba divers at a sound source should have a population mean of zero. In fact, the mean was $0.8°$ to the left of center. We tested this deviation using the standard error of the mean $s_x/n^{1/2}$ and with an s_x of $58°$ showed that the sample mean might lie anywhere within $\pm 3°$ of zero with 68% probability. If we apply the t statistic to this deviation,

$$t = \frac{(0.8 - 0.0)}{58} (350)^{1/2} = 0.259 \qquad \text{and} \qquad df = 350 - 1 = 349$$

Figure 9-5, which is set up for the *two-tailed test*, shows that this is a very probable value of t, to be expected much more often than 10% of the time. Since our hypothesis is that there is *no* difference between $\bar{X}$ and μ_X, we have *not disproved* it and conclude that the two are drawn from the same population. We might then ask, How large a difference $\bar{X} - \mu_X$ would we need to reject this hypothesis at the 5% level? With $df = 349$, Fig. 9-5 shows the 5% line asymptotic to a t value of 1.96. Then

$$\bar{X} - \mu = \frac{1.96(58)}{n^{1/2}} = 6.08°$$

Notice that two standard deviations of the mean also equal about $6°$, and thus $\pm 6°$ should enclose 95% and shut out 5%. In fact, with samples this large, the t distribution approaches the form of the normal distribution and gives the same result, as here. With samples of less than 30 items, it is better to test against the known mean with the t statistic than to use the standard error of the mean.

A more common use of the t statistic involves the hypotheses, "The means from two samples *having the same population SDs* come from the same population." For this case, t is found from

$$t = \frac{\bar{X}_a - \bar{X}_b}{s_{\text{total}} (1/n_a + 1/n_b)^{1/2}} \qquad\qquad (9\text{-}7)$$

where $\bar{X}_a$ is the mean of sample a of number of items n_a, $\bar{X}_b$ is the mean of sample b and n_b items, and s_{total} is the so-called combined standard deviation,

$$s_{\text{total}} = \left[\frac{s_a^2(n_a - 1) + s_b^2(n_b - 1)}{n_a + n_b - 2} \right]^{1/2} \qquad\qquad (9\text{-}8)$$

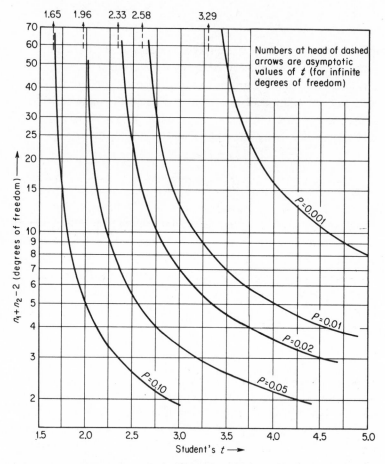

Figure 9-5 A plot showing the relationship between Student's t, the degree-of-freedom term $n_1 + n_2 - 2$, and the probability of obtaining a given value of t (or a larger one) if both data sets come from the same population.

and the degrees of freedom are found from

$$df = n_a + n_b - 2 \tag{9-9}$$

Example 9-4 In Example 8-1 we studied the statistical characteristics of two data sets relating to the hand- and foot-path resistance values through the human body. After rejecting two outlying points, we obtained a mean of 18,860 Ω with a SD of 5,717 Ω for the hand data and a mean of 15,272 Ω with a SD of 7,028 Ω for the foot-path data. We might expect the foot data to be lower because there is more contact area, but have we shown this?

Solution $n_a = n_b = 13$ points. Then Eq. 9-8 gives

$$s_{total} = \left[\frac{5{,}717^2(13 - 1) + 7{,}028^2(13 - 1)}{13 + 13 - 2} \right]^{1/2} = 6{,}405 \ \Omega$$

and Eq. 9-7 gives

$$t = \frac{18{,}860 - 15{,}272}{6{,}405 \ [(1/13) + (1/13)]^{1/2}} = 1.425$$

with $df = 26 - 2$, or 24. Figure 9-5 shows that we are well above a 10% probability, and the hypothesis that the two have the same means is *not* rejected. Thus we can say that as far as the test of the means, the two sets of data are *not* different and to detect any difference, if in fact one exists, would require more testing with more subjects. To obtain a significant value of $\bar{X}_a - \bar{X}_b$ with $df = 24$, we need a t of about 2.1 at the 5% level on Fig. 9-5. Thus, with s_{total} of 6,405 Ω,

$$\bar{X}_a - \bar{X}_b = 2.1(6{,}405)\left(\frac{1}{13} + \frac{1}{13} \right)^{1/2} = 5{,}275 \ \Omega$$

This difference is almost one-and-a-half times that found experimentally.

To accomplish this test, we assumed that the two SD values were drawn from a common population of values such that $\sigma_a = \sigma_b$. This is reasonable to expect in this body-resistance data because the SDs do not differ markedly. The t test can be applied to data sets that do not meet this requirement, but we shall not discuss this more complicated situation. In the next section we shall show how the F-ratio test can be used to test for the $\sigma_a = \sigma_b$ condition.

9-5 THE ANALYSIS OF VARIANCE

The *F distribution*, discovered by the British statistician, R. A. Fisher, gives us a final tool to make comparisons between two or more distributions of a random variable, as well as a means of analyzing *factorial plans* of the sort described in Chap. 6. We define the *F ratio* as

$$F = \frac{s_1^2}{s_2^2} \tag{9-10}$$

Table 9-1 F ratio for 0.05 probability

n_2	1	2	3	4	5	6	12	24	∞
1	164	200	216	225	230	234	235	249	254
2	18.5	19.2	19.2	19.3	19.3	19.3	19.4	19.5	19.5
3	10.1	9.6	9.3	9.1	9.0	8.9	8.7	8.6	8.5
4	7.7	6.9	6.6	6.4	6.3	6.2	5.9	5.8	5.6
5	6.6	5.8	5.4	5.2	5.1	5.0	4.7	4.5	4.4
6	6.0	5.1	4.8	4.5	4.4	4.3	4.0	3.8	3.7
8	5.3	4.5	4.1	3.8	3.7	3.6	3.3	3.1	2.9
10	5.0	4.1	3.7	3.5	3.3	3.2	2.9	2.7	2.5
12	4.8	3.9	3.5	3.3	3.1	3.0	2.7	2.5	2.3
16	4.5	3.6	3.2	3.0	2.9	2.7	2.4	2.2	2.0
20	4.4	3.5	3.1	2.9	2.9	2.6	2.3	2.1	1.8
30	4.2	3.3	2.9	2.7	2.5	2.4	2.1	1.9	1.6
60	4.0	3.2	2.8	2.5	2.4	2.3	1.9	1.7	1.4
∞	3.8	3.0	2.6	2.4	2.2	2.1	1.8	1.5	1.0

where $s_1 > s_2$. Also, the distribution has two different degrees of freedom,

$$(df)_1 = n_1 - 1$$

$$(df)_2 = n_2 - 1$$

(9-11)

Table 9-1 gives the F value that divides the F distribution into a 5% tail and a 95% body. If the experimental value is higher than the tabulated one, we may reject the hypothesis of no difference at the 5% level. This test has a similar restriction to that required with the t test, only now we must use data sets having a *common population mean*, $\mu_1 = \mu_2$. Thus we use the t statistic when we think that the dispersions of the two samples are similar and the F-ratio test when we think that the means are similar. If both are *known* to be different, a statistics text should be consulted, or we can fall back on the use of the *standard error of the mean* described in Sec. 2-7, Eq. 2-25.

Example 9-5 We saw in Example 9-4 that the means of the body-resistance data were not significantly different. What about the SDs?

Solution We form the ratios of the variances:

$$F = \frac{7028^2}{5717^2} = 1.511 \qquad \text{and} \qquad (df)_1 = 12 = (df)_2$$

The hypothesis is that there is no difference between the means. To reject this at the 5% level, we need an F of 2.7 from Table 9-1, considerably larger than the experimental 1.511. To obtain this large an F, the ratio of s_1/s_2 would have to be about 1.64 instead of 1.23. Alternatively, we would need well in excess of 60 data with this large a difference to challenge the hypothesis.

We see finally, then, that the two samples of body-resistance data are in *no way statistically different.* The foot path is *not* proven to be lower in resistance, and its dispersion is not larger. In fact, it would be reasonable to *pool* the data and treat them from now on as a single 26-item sample.

In Sec. 6-6 we saw how several variables could be studied at once using the so-called factorial plan. Example 6-3 showed how such a plan could be applied to data of *high precision.* In fact, we saw in Example 7-3 that the ambient temperature of the test was discoverable from those data.

Factorial plans are more often used with data of *low precision,* where we apply several different *treatments* at several *levels* and ask if any treatment has statistically significant effects. Although such plans can be in any rectangular format, with or without replications, we shall restrict this study to plans that are *balanced,* that is, in which *all the variables* are changed over the *same number of levels*, as in the Graeco-Latin square plans shown in Chap. 6.

If we imagine such a plan, $n \times n$ levels on a side, with results X_1, $\ldots, X_{n \times n}$, we define the *total variation* v by

$$v = \sum_{1}^{n \times n} (X - \bar{X})^2 \tag{9-12}$$

A shortcut for finding v is

$$v = \sum_{1}^{n \times n} X^2 - \frac{\left(\sum_{1}^{n \times n} X \right)^2}{n^2} \tag{9-13}$$

We sort the data into n groups based on the n levels of treatment A and define the *variation due to A, v_a,* from

$$v_a = n \sum_{1}^{n} (\bar{X}_a - \bar{X})^2 \tag{9-14}$$

where $\bar{X}_a$ is the separate mean of the results for each level of treatment A. If we had, for example, a 4×4 square plan, we would have four sets of four numbers, one set for each A level. We then find these four means and compute v_a from Eq. 9-14. A shortcut is

$$v_a = \frac{1}{n} \sum_1^n \left(\sum_1^n X_a \right)^2 - \frac{\left(\sum_1^{n \times n} X \right)^2}{n^2} \tag{9-15}$$

where we simply sum all the A-treatment, level 1 results and square, then all the A-treatment, level 2 results and square, and finally sum the n sums of these squares.

If we have other treatments, B, C, and so on, we use identical equations to find v_b, v_c, and so on. We could have as many as four treatments in a Graeco-Latin square.

We assume that the total variation found by Eq. 9-12 or 9-13 is made up of all the treatment-produced variations v_a, v_b, and so on, plus the *variation due to error*, v_{error}. Thus

$$v_{\text{error}} = v - v_a + v_b + v_c + v_d \tag{9-16}$$

where Eq. 9-16 assumes a four-treatment experiment.

If we then form the variance of the variation due to A, v_a, from

$$s_a^2 = \frac{v_a}{n-1} \tag{9-17}$$

and similar variances due to B, C, and D, plus error, we can then compare them using the F statistic and make decisions about which treatments have the greatest effect. In a balanced experimental design, the denominator of Eq. 9-17 will always be $n-1$ for all treatments. For the v_{error} variance, however, we form s_{error}^2 from

$$s_{\text{error}}^2 = \frac{v_{\text{error}}}{(n-1)(n+1-N_{tr})} \tag{9-18}$$

and the degrees of freedom are simply these denominators,

$$(df)_a = n - 1$$

$$(df)_{\text{error}} = (n-1)(n+1-N_{tr}) \tag{9-19}$$

where N_{tr} is the number of treatments used, with a usual maximum of four. If, for example, we had a 5×5 Latin square, n is 5 and N_{tr} is 3. Then $(df)_a$ is 4 for each of the three treatment variances, and $(4)(6-3)$, or 12, for the error variance. The sum of the N_{tr} treatment dfs and the error df must equal $n^2 - 1$.

Example 9-6 Figure 9-6 shows a test in which 64 "identical" electrical resistors are subjected to the same overload voltage and the time that each burns out is recorded. Four treatments of the resistors were tested: The resistors, all nominally of 50 Ω, were segregated into four groups, 48–49, 50–51, 52–53, and 54–55 Ω, none lying outside this range. Four different mounting devices were used, M_1 through M_4; glass covers were used to simulate a resistor at the bottom of a chassis (on or off); and the location by column was assumed to be a treatment, because there was a cold window at the left side of the test area. Table 9-2 shows the Graeco-Latin square test plan. Note that the covering treatment is randomized over half the plan, because we have only two levels of this treatment. Table 9-3 shows the raw data resulting from the test. What can we say about the effects of the several treatments on overload life?

Solution Although the fact that we replicated each test four times adds to the ways we might analyze the test, we shall simply assume that the mean of

Figure 9-6 Photo of a factorial experiment on resistor failure, as described in Example 9-6.

Table 9-2 Graeco-Latin square test plan

Resistance value	Location 1	Location 2	Location 3	Location 4
R_1, 48–49 Ω	M_1, on	M_2, on	M_3, off	M_4, off
R_2, 50–51 Ω	M_4, off	M_3, off	M_2, on	M_1, on
R_3, 52–53 Ω	M_2, on	M_1, off	M_4, off	M_3, on
R_4, 54–55 Ω	M_3, off	M_4, on	M_1, on	M_2, off

each of the 16 positions represents the X variable at that place. Then Fig. 9-7 shows the results and the *treatment means* as well. We note that the covering effect is clearly very small, because the two means are almost the same. We shall thus include the covering-treatment effect in the variation due to error, treating this as a Latin square with N_{tr} of 3. We first need the total variation v from Eq. 9-13:

$$v = 94.77^2 + 96.8^2 + 99.3^2 + \cdots + 120.94^2$$

$$- \frac{(94.77 + 96.8 + 99.3 + \cdots + 120.94)^2}{16}$$

$$= 174{,}493 - 173{,}075 = 1{,}418 \text{ min}^2$$

Next we consider the effect of the different resistance values, which in the Table 9-2 plan are varied by row. From Eq. 9-15,

$$v_{res} = \frac{(98.45 \times 4)^2 + (94.02 \times 4)^2 + (109.94 \times 4)^2 + (113.51 \times 4)^2}{4}$$

$$- 173{,}073$$

$$= 174{,}090 - 173{,}073 = 1{,}016 \text{ min}^2$$

Table 9-3 Minutes and seconds to failure[a]

26:20	46:30	41:35	28:18	34:45	38:15	47:45	42:00
36:20	29:55	43:15	34:05	38:20	46:00	42:15	46:00
26:35	28:05	21:30	38:00	31:20	38:40	37:15	53:30
32:15	33:40	37:00	20:20	38:50	18:10	42:45	46:00
39:45	62:45	32:20	39:00	53:45	47:00	55:50	43:20
52:25	52:29	56:20	36:45	70:45	30:40	52:00	59:30
27:30	71:45	63:30	50:30	69:45	39:00	68:30	66:30
48:30	56:30	57:50	33:30	46:35	61:45	47:00	61:45

[a]Add 1 h to each value for total lifetime.

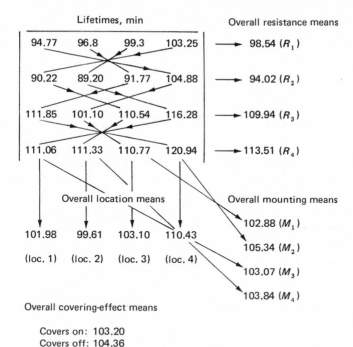

Overall covering-effect means

Covers on: 103.20
Covers off: 104.36

Figure 9-7 Results and treatment means for Example 9-6.

The location treatment varies by *column*. We can find the sum of the squares of the columns and use Eq. 9-15 to obtain $v_{\text{loc}} = 306$ min^2. Similarly, for the mounting treatment, which is randomized within the square, we obtain $v_{\text{mount}} = 6.0$ min^2. Then Eq. 9-16 gives

$$v_{\text{error}} = 1418 - 1016 + 306 + 6 = 90 \text{ min}^2$$

To decide if resistance value has an effect on lifetime, we form the usual null hypothesis, that the resistance treatment has no more effect on the variance than random, that is, error effects. Let us check this using the ratio of the variance resulting from the treatment to the variance due to error: From Eq. 9-17,

$$s_{\text{res}}^2 = \frac{1016}{4 - 1} = 339 \text{ min}^2$$

Equation 9-18 gives the error variance,

$$s_{\text{error}}^2 = \frac{90}{(4 - 1)(4 + 1 - 3)} = 15 \text{ min}^2$$

Note that $(df)_{res}$ is $4-1$, or 3, and $(df)_{error}$ is $(3)(2)$, or 6. We can now form the F ratio,

$$F_{res} = \frac{339}{15} = 22.6$$

Entering Table 9-1 with n_1 of 3 and n_2 of 6, we note that $F_{3,6}$ must exceed 4.8 to reject our hypothesis at the 5% level. Clearly, we are way beyond this. The effect of resistance is of great importance in determining which resistors burn out first. Of course we should expect that, since power is E^2/R, with a constant voltage E, the resistor must dissipate more heat as R decreases.

The complete results of this balanced-square study are often given in an *analysis-of-variance table*:

Variation	df (Eq. 9-19)	Variance or mean square	F (Eq. 9-10)	$F_{5\%,3,6}$ (Table 9-1)
Resistance (rows)	3	339 (Eq. 9-17)	$\frac{339}{15} = 22.6$	4.8
Location (columns)	3	102 (Eq. 9-17)	$\frac{102}{15} = 6.8$	4.8
Mounting (in square)	3	2.0 (Eq. 9-17)	$\frac{2}{15} = 0.13$	4.8
Error (Eq. 9-16)	6	15 (Eq. 9-18)		
Total	15			

We see that resistance variation is the decisive treatment but that the location variance is also significant at the 5% level. Since location is an extraneous or undesirable variation, we should move the test location to a thermally more uniform area or else baffle it. The mountings have no effect on life and that treatment could be simply ignored in future tests, any of the four mountings being used interchangeably.

Balanced experiments with no replication have the advantage of relatively straightforward and uncomplicated equations and rules for finding degrees of freedom, but they lead to a rather restricted set of experimental plans. We note in Eq. 9-19 that if N_{tr} is 4 (Graeco-Latin plan) and n is 3, then $(df)_{error}$ will be zero. Evidently we cannot investigate the effect of four treatments in a 3×3

Table 9-4 Degrees of freedom of error

n	Complete block $N_{tr} = 2$	Latin square $N_{tr} = 3$	Graeco-Latin square $N_{tr} = 4$
2	1	–	–
3	4	2	–
4	9	6	3
5	16	12	8
6	25	20	Impossible for 6 × 6

square. Table 9-4 shows the value of this important parameter (*df* of error) for the plans most likely to prove possible. Clearly, 2 × 2 squares offer very little. The degree of freedom of error of 1.0 requires huge differences in variances, as Table 9-1 shows. The 3 × 3 square is really suitable only for two-treament experiments. As Latin squares, the $F_{5\%,2,2}$ of 19.2 is still very large and would require very strong effects to disprove the hypothesis. The most practical plans for the bulk of engineering tests are probably 4 × 4 squares. They require only 16 runs but, as we saw, can detect reasonably subtle effects (such as the location in Example 9-6). The 5 × 5 squares offer much more powerful testing with an increase of only nine more runs, but there is no 6 × 6 Graeco-Latin square. The factorial, balanced-test planner is thus really confronted with either a four-level or a five-level experiment in almost all cases.

9-6 INSTRUCTIONS FOR USING CHAPTER 9

Chapter 9 is reached through two routes. Either the data are progressively shown to require pure statistical methods, as in the electrical-resistance data discussed in Examples 8-1, 8-4, 9-4, and 9-5, or the experiment is laid out to give data requiring statistical conclusions from the beginning. In the case of the underwater sound location tests reported in Example 9-1, there was never any doubt that the data would be so random as to require statistical conclusions.

If the data are suspected to be purely random, they are often compared either to the normal equation as in Sec. 2-10 or to the Poisson equation, which starts at zero, as in Sec. 9-3. This comparison is made with numbers of data using the χ^2 test (Sec. 9-2), and we must decide which kind of error we wish to risk (Sec. 9-1) and then choose a probability limit beyond which we say that our hypothesis of no difference is disproven.

When we have continuous distributions of data, we can compare their means, using the t test (Sec. 9-4), or their variances, using the F-ratio test (Sec. 9-5), but in each case we must assume that the other quantity is the same for the two populations. The t test also does for small samples what the standard deviation of the mean in Sec. 2-7, Eq. 2-25, does for large: It allows us to compare a sample mean with a population mean or fixed value.

Factorial tests, in which up to four variables can be investigated simultaneously with regard to their effect on a fifth, were described and the plans shown in Chap. 6. Using the analysis of variance in Sec. 9-5, the results of such tests can be analyzed and the relative importance of each variable on the output assessed. However, such testing in square form is really restricted to four- and five-level experiments.

PROBLEMS

9-1 Decide whether you would rather risk a type 1 or a type 2 error in the following tests of hypothesis, and give a one-sentence justification of your choice.

(*a*) The hypothesis is "There is no greater incidence of leukemia in children living in or out of the heavy fallout belt."

(*b*) The hypothesis is "Over a 3-year period of test there is no difference in repair costs between an outboard motor with a bronze journal bearing and one with ball bearings."

(*c*) The hypothesis is "The addition of seat belts to automobiles makes no difference in survival in a crash."

(*d*) The hypothesis is "Flying saucer sightings are not more frequent during radar-to-Venus experiments."

9-2 Four methods, *A, B, C,* and *D,* were applied to the fabrication of a difficult machined part. The following rejection rates occurred:

	Method			
	A	*B*	*C*	*D*
Total parts made	8	10	9	13
Number rejected	5	8	9	10

Test the hypothesis that there is no difference in rejection rate among the methods.

9-3 During a semester's work in a fluid mechanics lab, the Monday group blew the mercury out of nine manometers, the Tuesday group blew six manometers, and the Wednesday group blew two. Test the hypothesis that there is no difference in lab ability among the three groups.

9-4 Three shifts, *A, B,* and *C,* are producing identical units. In the first day, the following results occur:

Shift	*A*	*B*	*C*
Acceptable units	6	6	9

Test the hypothesis that there is no difference among the three shifts. We suspect that shift C is the best one. Suppose that the same results as above occurred in each succeeding shift. How many shifts need we wait before this hypothesis is disproved to a 5% level? To a 1% level?

9-5 Twenty specimens of the same kind are tested on a tension test machine to failure, 10 at a fast rate and 10 at a slow loading rate. Seven of the slow specimens show a complete cup-cone fracture, with the other three showing a partial cup. Four of the fast specimens show a complete cup cone, with six showing a partial one only. Test the appropriate hypothesis.

9-6 Two physicists count the number of delta rays per 100 μm on the same photographic plate as follows:

A counts	10	23	9	46	7	11	10	15	7	8	12	36	28
B counts	8	21	8	43	7	11	10	12	6	8	11	35	29
A − B	2	2	1	3	0	0	0	3	1	0	1	1	−1

A similar series of counts is made on a second plate, and the absolute value of $A - B$ becomes (17 readings) 0, 0, 1, 2, 0, 1, 0, 0, 3, 1, 1, 0, 1, 6, 2, 0, 2. Using the t test, compare these two sets of deviations, and decide if they come from the same infinite population.[*]

9-7 In the years 1893 and 1894, Lord Rayleigh extracted nitrogen from air, using both iron and copper, and from nitrous oxide, again using iron. The resulting densities of nitrogen from the air were 2.30986, 2,31001, 2.31010, 2.31017, 2.31012, 2.31024, 2.31026, 2.31027, 2.31035; and from the experiment using nitric oxide, 2.29816, 2.29890, 2.31043, 2.30182, 2.29869, 2.29940, 2.29849, 2.29889. From these data Lord Rayleigh concluded that the atmospheric nitrogen was actually heavier than the chemically produced nitrogen, and he was thus led to the discovery of the rare gases. Decide if he was statistically correct in this decision, using the t test.

9-8 Repeat Prob. 9-7 but use the χ^2 test to test the hypothesis: "Taking the mean of all 17 determinations, there is no difference in the direction of deviation between the atmosphere and chemical methods." An alternative statement might be: "There is no difference in the two methods as to whether the given determination falls above or below the mean value."

9-9 The speed of a shaft is found by an absolute method (oscilloscope and signal generator calibrated against the line frequency), and a value of 1010 rpm results. A strobotac and a hand tachometer give

Strobe	1000	980	995	1020	1005
Tach	990	1020	1000	1010	1040

Test the following hypotheses in turn: "The two series of readings are from the same population of readings." "The deviations of these sets of readings from the true value are from the same population of such deviations." For this second hypothesis, you should obtain your averages of the deviations with regard to sign.

9-10 Metal-and-glass-enclosed radio tubes of the same type are life tested at extreme conditions. The results are

[*]*Am. J. Phys.*, vol. 24, no. 3, pp. 157–159, March, 1956.

Metal tube, h	53	40	92	67	89
Glass tube, h	45	40	47		

Use the t test to check the appropriate hypothesis.

9-11 Concrete taken from a batch mixed on May 25 gives compression strengths of 4346, 3852, 4240, 3109, 3887, 3852, 3263, and 3781, all in psi. The material mixed on June 17 gives strengths of 4240, 3746, 4099, 4134, 4664, 4311, 3956, 4211, 4488, 4134, 4523, 3852, 4346, 4558, 4170, 4060, and 4488, all in psi. Using the t and F-ratio tests, test whether the two batches of mix were uniform as to (a) means, and (b) variances.[*]

9-12 Test the following distribution(s), using a Poisson plot and/or a χ^2 test:

(a) Student's data on the number of yeast cells in given areas:

Cells/area, n	0	1	2	3	4	5	6	7	8	9	10	11	12
Areas with n cells	0	20	43	53	86	70	54	37	18	10	5	2	2

(b) Number of defectives in 51 shifts:

Number/shift, n	0	1	2	3	4	5	6	7	8
Shifts with n defectives	3	7	9	12	9	6	3	2	0

(c) Intervals between cosmic-ray counts using a cosmic-ray counting array and paper-tape recording:

Paper marks per fixed interval, n	0	1	2	3	4	5	6	7	8
Intervals having n marks	2	33	182	333	318	194	70	17	4

9-13 The following data report an air-pollution survey over a period of 210 days at Crescent City, California. The average oxidant concentration during the day in parts per million is tabulated with the number of days that such a concentration occurred:

Oxidant concentration, ppm $\times 10^2$	0	1	2	3	4	5	6	7	8 or more
No. of days with concentration shown	1	5	28	19	11	7	1	1	0

Analyze these data for similarity to a Poisson distribution.

9-14 Out of a total 100 six-month periods, 32 storms having force 7 or higher winds are recorded at a location on the Atlantic Ocean. Further, the distribution of such storms throughout the 100 periods is a Poisson one. The maximum waves generated by such storms are normally distributed in height with a mean value of 8 ft and a modulus of precision of 0.25 ft^{-1}. An offshore drilling rig designer estimates that waves less than 13 ft high will not damage drilling equipment. It is further estimated that the structure will not be totally lost

[*]*J. Am. Concrete Inst.,* vol. 26, no. 8, pp. 765–772, 1955.

unless it experiences wave damage at least twice in a 6-month period (longer times permit repairs). What is the chance of having two or more storms within a given 6-month period? What is the chance of a storm having damaging (13-ft) waves? What is the probability of a given tower collapsing completely in a 10-year period?

9-15 Three different sets of relays, each having the same number N, are to be tested for a million-cycle life test. If all of groups A and B fail and three-fourths of group C fails, what must be the value of N to establish C as superior at the 5% confidence level?

9-16 Two instruments, A and B, are used to give thermal conductivity of a given sample:

A, Btu/h · °F · ft	9.56	9.49	9.62	9.51	9.58	9.63	
B, Btu/h · °F · ft	9.33	9.31	9.47	9.01	9.56	9.0	9.47

What can we say about the *variability* of these two instruments? Which is preferable from this standpoint, and would we be justified in using it exclusively?

9-17 A study of wrong numbers in a telephone system reveals the following data:

N	0	1	2	3	4	5	6	7	8	9	10	11	12	13	14	15	16
P	0	0	1	5	11	14	22	43	31	40	35	20	18	12	7	6	2

Decide if a Poisson standard distribution fits the data when N is the number of wrong numbers and P is the number of periods exhibiting the given N.

9-18 At a given station, the number of storms is counted in a year. Over a period of several years and including data from a number of stations, the following results:

Storms per station per year	0	1	2	3	4	5	6 or more
No. of such occurrences	102	114	74	28	10	2	0

Decide if these actual data[*] could be approximated by a Poisson distribution.

9-19 An engineer tests 32 "identical" metallurgical samples. He notes that seven out of the first eight fail, then numbers 11 and 12, then none until numbers 17, 19, 27, 30, and 32. He believes that the test equipment or test methods underwent a considerable alteration at some point during the test. Decide whether it is more likely that this change occurred on specimen 9 or on number 13.

9-20 The following data were compiled in the state of Connecticut during the period 1931-1936 regarding the number of accidents any given driver in a random sample might have and the number of drivers during the period having this number:

Accidents per operator	0	1	2	3	4	5	6	7
No. of operators with this many accidents	23,881	4,503	936	160	33	14	3	1

What do you conclude about this sample of drivers from these actual data?[†]

[*] E. L. Grant, *Trans. Am. Soc. Civil Engrs.*, vol. 103, pp. 384–388, 1938.
[†] C. Johnson, *Proc. Highway Res. Board,* Washington, D.C., 1937, pp. 444–454.

9-21 A square plan identical to that shown in Table 9-2 is run with a similar group of resistors but with a higher overload voltage. The following mean lifetimes occur:

Resistor lifetimes (min)			
43.65	48.2	47.2	50.00
42.2	39.9	43.7	39.8
36.2	39.3	35.7	36.1
38.3	36.1	37.9	41.1

Analyze this result using the methods of the analysis of variance. Discuss the relative importance of the resistance, location, and mounting treatments but keep the covering treatment in the error, as in the text example. Does this experiment agree with the findings of the text example?

9-22 The useful lives of batteries are tested at a given load, the units being from three different manufacturers and being tested at three different temperatures.

Temperature	Battery lives (h)		
	Type A	Type B	Type C
50°C	75	78	80
40°C	81	76	79
30°C	73	75	77

Using the analysis-of-variance method, decide if either or both treatments have significant effects on the variations observed.

9-23 In Example 3-1 we noted a device to obtain the kinetic coefficient of friction μ. This measurement was made at a given place using the pendulum device described in Example 3-1, a weighted box, and an automobile.

	Trial							
	1	2	3	4	5	6	7	8
μ, pendulum	0.89	0.84	0.85	0.85	0.80	0.83	0.79	0.90
μ, box	0.88	0.88	0.86	0.84	0.87			
μ, car	0.87	0.88	0.88	0.88				

Discuss whether these three data sets come from the same (*a*) population of means, and (*b*) population of variances. Assuming the auto data to be what we wish to predict, decide which instrument is preferable.

9-24 If the standard deviations (min) of the Table 9-2 resistor test plan are as shown,

8.85	6.93	4.48	1.86
3.29	9.60	9.68	6.79
9.42	10.52	16.60	6.94
18.53	13.02	13.23	9.71

we can use the analysis of variance method to study the effect of the several treatments on variability of the resistor lifetimes. Perform this analysis and discuss the results.

9-25 Referring to the body-resistance data of Prob. 8-9, decide whether either the means or the SDs come from different populations but be certain that all point rejection has been done first, as Prob. 8-9 suggests.

9-26 In Example 9-1 we studied whether the diver-pointing data from 800-Hz tests were the same as data from 2000- and 4000-Hz tests using a success–failure approach and the χ^2 statistic. The F-ratio test offers a more powerful test, because it uses the dispersion of the points, not just where they are in relation to a single value. Also, we have seen that all pointing trials have a mean close to zero. The 36 800-Hz tests gave a SD of 37.6°, and the 49 high-frequency tests had a SD of 65.6°. Does the F-ratio test confirm our previous conclusion that these are data from different populations?

9-27 Example 8-1 and Prob. 8-9 give two data sets relating to hand–hand and hand–foot electrical resistance in human subjects. Are the two sets of hand–hand sets from the same population of (a) means, and (b) variances? Decide this for the hand–foot data sets as well.

BIBLIOGRAPHY

Bennett, C. A., and N. L. Franklin: *Statistical Analysis in Chemistry and the Chemical Industry,* Wiley, New York, 1954.

Brownlee, K. A.: *Industrial Experimentation,* Chemical, New York, 1953.

Cramer, H.: *The Elements of Probability Theory,* Wiley, New York, 1955.

Fisher, R. A.: *Statistical Methods for Research Workers,* Hafner, New York, 1954.

Greenshields, B. W., and F. M. Weida: *Statistics with Application to Highway Traffic Analysis,* Eno Foundation, Saugatuck, Conn., 1952.

Greenwood, M., and G. Yule: An Inquiry into the Nature of Frequency Distributions Representative of Multiple Happenings, *J. Roy. Stat. Soc.,* vol. 83, pp. 255–279, 1920.

Guest, P. G., and W. M. Simmons: An Experiment on Cosmic Rays, *Am. J. Phys.,* vol. 21, no. 5, pp. 362–367, May, 1953.

Hald, A.: *Statistical Theory with Engineering Applications,* Wiley, New York, 1952.

International Business Machines: *Call/360: Statistical Package (Statpack) Version 2,* International Business Machines, White Plains, N.Y., 1970.

Mood, A. M., and F. A. Graybill: *Introduction to the Theory of Statistics,* 2nd ed., McGraw-Hill, New York, 1963.

Parratt, L. G.: *Probability and Experimental Errors in Science,* Wiley, New York, 1961.

Schenck, H.: An Accelerated Life Test Using a Graeco-Latin-Square Test Plan, *Bull. Mech. Eng. Educ.,* vol. 3, pp. 241–245, July–September, 1964.

Spiegel, M. R.: *Probability and Statistics,* Schaum's Outline Series, McGraw-Hill, New York, 1975.

Tippett, L. H.: *Technological Applications of Statistics,* Wiley, New York, 1950.

Volk, William: *Applied Statistics for Engineers,* McGraw-Hill, New York, 1958.

Wilson, E. B.: *An Introduction to Scientific Research,* chap. 8, McGraw-Hill, New York, 1952.

CONVERSION FACTORS TO SI UNITS

Quantity	Conversion	Multiply by
Length	in to m	0.0254
	ft to m	0.3048
	miles to m	1609.344
Area	in^2 to m^2	0.0006451
	ft^2 to m^2	0.09290
Volume	in^3 to m^3	0.000016387
	ft^3 to m^3	0.028317
	liters to m^3	0.001
Mass	lb (avoir) to kg	0.45359
	slug to kg	14.5939
Acceleration	ft/s^2 to m/s^2	0.3048
	Standard gravity (m/s^2)	(equal to) 9.80665
Force	lb_f to N	4.4482
Pressure—stress	lb_f/in^2 to Pa	6894.76
Energy—work	Btu (IT) to J	1055.06
	ft · lb_f to J	1.35582
Power	hp to W	745.699
Temperature	°F to °C	$t_C = (t_F - 32.)/1.8$
	°F to K	$t_K = (t_F + 459.67)/1.8$
Temperature interval	°F or °R to K or °C	0.555556

DIMENSIONAL FORMULAS

M = mass, L = length, θ = time, T = temperature, H = heat, K = dielectric constant, μ = magnetic field constant.

Mechanics

Quantity	Dimensional formula
Length L	L
Volume V	L^3
Curvature	L^{-1}
Velocity V	$L\theta^{-1}$
Acceleration A or g	$L\theta^{-2}$
Angular velocity ω	θ^{-1}
Density ρ	ML^{-3}
Momentum	$ML\theta^{-1}$
Angular momentum	$ML^2\theta^{-1}$
Force F	$ML\theta^{-2}$
Work and energy	$ML^2\theta^{-2}$
Power	$ML^2\theta^{-3}$
Viscosity μ	$ML^{-1}\theta^{-1}$
Kinematic viscosity	$L^2\theta^{-1}$
Surface tension	$M\theta^{-2}$
Pressure P	$ML^{-1}\theta^{-2}$

Thermal quantities

Quantity	Thermal formula	Dynamical formula
Quantity of heat H	H	$ML^2\theta^{-2}$
Specific heat C_p	$HM^{-1}T^{-1}$	$L^2\theta^{-2}T^{-1}$
Thermal conductivity k	$HL^{-1}\theta^{-1}T^{-1}$	$LM\theta^{-3}T^{-1}$
Heat-transfer coefficient h	$HL^{-2}\theta^{-1}T^{-1}$	$M\theta^{-3}T^{-1}$
Entropy s	HT^{-1}	$ML^2\theta^{-2}T^{-1}$
Coefficient of thermal expansion β	T^{-1}	T^{-1}

Magnetic and electrical quantities

Quantity	Electromagnetic	Electrostatic
Magnetic field strength	$M^{1/2}L^{-1/2}\theta^{-1}\mu^{-1/2}$	$M^{1/2}L^{1/2}\theta^{-2}K^{1/2}$
Magnetic pole strength	$M^{1/2}L^{3/2}\theta^{-1}\mu^{1/2}$	$M^{1/2}L^{1/2}K^{-1/2}$
Electric current	$M^{1/2}L^{1/2}\theta^{-1}\mu^{-1/2}$	$M^{1/2}L^{3/2}\theta^{-2}K^{1/2}$
Quantity of electricity	$M^{1/2}L^{1/2}\mu^{-1/2}$	$M^{1/2}L^{3/2}\theta^{-1}K^{1/2}$
Potential difference	$M^{1/2}L^{3/2}\theta^{-2}\mu^{1/2}$	$L^{1/2}M^{1/2}\theta^{-1}K^{-1/2}$
Resistance	$L\theta^{-1}\mu$	$L^{-1}\theta K^{-1}$
Capacitance	$L^{-1}\theta^{+2}\mu^{-1}$	LK
Inductance	$L\mu$	$L^{-1}\theta^2K^{-1}$
Permeability	μ	$L^{-2}\theta^2K^{-1}$
Permittivity	$L^{-2}\theta^2\mu^{-1}$	K

GRAECO-LATIN SQUARES

The following designs may be used in the planning of randomized experiments as explained in Chaps. 6 and 9. Notice that each plan is actually two different Latin squares.

3 × 3

A1	B3	C2
B2	C1	A3
C3	A2	B1

4 × 4

A1	B3	C4	D2
D4	C2	B1	A3
B2	A4	D3	C1
C3	D1	A2	B4

5 × 5

A5	B3	C2	D1	E4
B1	C4	D5	E3	A2
C3	D2	E1	A4	B5
D4	E5	A3	B2	C1
E2	A1	B4	C5	D3

6 × 6 (only the Latin square possible)

A	B	C	D	E	F
B	F	D	C	A	E
C	D	E	F	B	A
D	A	F	E	C	B
E	C	A	B	F	D
F	E	B	A	D	C

INDEX